P9-DCU-375

2nd Edition

Genetically Modified Crops

2nd Edition

Genetically Modified Crops

Nigel G Halford

Rothamsted Research, UK

ICP

Imperial College Press

Published by

Imperial College Press
57 Shelton Street
Covent Garden
London WC2H 9HE

Distributed by

World Scientific Publishing Co. Pte. Ltd.
5 Toh Tuck Link, Singapore 596224
USA office: 27 Warren Street, Suite 401-402, Hackensack, NJ 07601
UK office: 57 Shelton Street, Covent Garden, London WC2H 9HE

British Library Cataloguing-in-Publication Data
A catalogue record for this book is available from the British Library.

GENETICALLY MODIFIED CROPS
2nd Edition

ISBN-13 978-1-84816-838-1
ISBN-10 1-84816-838-1

Typeset by Stallion Press
Email: enquiries@stallionpress.com

Printed by FuIsland Offset Printing (S) Pte Ltd Singapore

CONTENTS

Preface		ix
1.	**DNA, Genes, Genomes and Plant Breeding**	**1**
1.1	A Brief History of Genetics	1
1.2	Deoxyribonucleic Acid (DNA)	4
1.3	Genes	8
1.4	Gene Expression	9
1.5	Genomes	10
1.6	Genetic Change	11
1.7	Plant Breeding	11
1.8	Modern Plant Breeding	15
1.9	Wide and Forced Crossing and Embryo Rescue	18
1.10	Radiation and Chemical Mutagenesis	19
1.11	The Advent of Genetic Modification	20
2.	**The Techniques of Plant Genetic Modification**	**23**
2.1	A Brief History of the Development of Recombinant DNA Technology	23
2.2	*Agrobacterium tumefaciens*	26
2.3	Use of *Agrobacterium tumefaciens* in Plant Genetic Modification	27
2.4	Transformation of Protoplasts	30
2.5	Particle Gun	31

2.6	Other Direct Gene Transfer Methods	32
2.7	*Agrobacterium*-mediated Transformation Without Tissue Culture	34
2.8	Selectable Marker Genes	34
2.9	Visual/Scoreable Marker Genes	38
2.10	Design and Construction of Genes for Introduction into Plants	40
2.11	Promoter Types	43
2.12	The Use of GM to Characterise Gene Promoters	45
2.13	Gene Over-Expression and Silencing	47
3.	**The Use of GM Crops in Agriculture**	**51**
3.1	Why Use Genetic Modification (GM) in Plant Breeding?	51
3.2	Slow-ripening Fruit	55
3.3	Herbicide Tolerance	57
3.4	Insect Resistance	64
3.5	Virus Resistance	68
3.6	Modified Oil Content	70
3.7	Modified Starch for Industrial and Biofuel Uses	80
3.8	High Lysine Corn	83
3.9	Vitamin Content: Golden Rice	84
3.10	Fungal Resistance	88
3.11	Drought, Heat and Cold Tolerance; Climate Change	90
3.12	Salt Tolerance	94
3.13	Biopharming	97
3.14	Removal of Allergens	103
3.15	Conclusions	105
4.	**Legislation Covering GM Crops and Foods**	**107**
4.1	Safety of GM Plants Grown in Containment	107
4.2	Safety of Field Releases of GM Plants	111

4.3 Safety of GM Foods 115
4.4 European Union Regulations 118
4.5 Labelling and Traceability Regulations 120
4.6 Safety Assessment and Labelling Requirements in the USA 124

5. Issues that have Arisen in the GM Crop and Food Debate 127

5.1 Are GM Foods Safe? 133
5.2 Will Genetic Modification Produce New Food Allergens? 134
5.3 Is it Ethical to Transfer Genes Between Different Species? 136
5.4 Animal Studies 137
5.5 GM Crops 'Do Not Work' 138
5.6 Did Tryptophan Produced by Genetic Modification Kill People? 139
5.7 The Monarch Butterfly 141
5.8 The Pusztai Affair 142
5.9 Alarm Caused by Contradictory Results of Biosafety Studies 144
5.10 'Superweeds' 146
5.11 Insect Resistance to Bt Crops 147
5.12 Segregation of GM and non-GM Crops: Co-existence of GM and Organic Farming 148
5.13 Antibiotic Resistance Marker Genes 150
5.14 Patenting 152
5.15 Loss of Genetic Diversity 153
5.16 The Dominance of Multinational Companies 154
5.17 The StarLink and ProdiGene Affairs 155
5.18 The *Cauliflower mosaic virus* 35S RNA Gene Promoter 157
5.19 Implications for Developing Countries 158
5.20 'Terminator' Technology 160
5.21 Unintentional Releases 161

5.22 Asynchronous Approvals 163
5.23 The United Kingdom Farm-Scale Evaluations 163
5.24 Conclusions 165

Index 169

PREFACE

It is fifteen years since the first large-scale cultivation of genetically modified (GM) crops and eight years since the first edition of this book was published. Since then, the use of GM crops has continued to rise around the world, with 134 million hectares being grown in 2009. The number of traits that have been introduced by genetic modification remains limited and there is still no established market for GM varieties of some of the world's major crops, namely rice, wheat and potato. GM rice looks set to take off but GM wheat remains some way off and GM potato, after some false starts in the 1990s, is currently limited to the early stages of development of a variety designed to produce industrial starch. Nevertheless, predictions that the technology would be lost in the face of fierce opposition from pressure groups and consumer hostility in some parts of the world have proved incorrect and GM crop varieties continue to be popular with farmers wherever they have been made available.

For plant scientists in the United Kingdom and the rest of Europe the situation remains a frustrating one. The GM crop debate of the late 1990s and early 2000s was lost and great damage was done to the European plant biotechnology industry as a result. The European regulatory system is now so difficult to negotiate that biotechnology companies are focused on obtaining approval for their products to be imported into Europe so that they can be grown elsewhere rather than approval for their varieties to be cultivated in Europe. European farmers are therefore becoming increasingly disadvantaged as they have to compete with GM crop varieties that they are not allowed to grow. As non-GM crop products become more difficult to source it is recognised that the situation cannot continue as it is, but efforts to re-start the debate are greeted with the same shrill predictions of disaster and the same headlines as before. Meanwhile, the USA, Argentina, Brazil, India, Canada, China, Paraguay, South Africa and 17 other countries all extend their lead in developing what will be a key twenty-first century technology.

In this edition, the chapter on the use of GM crops in agriculture is expanded while those covering techniques, legislation and the GM crop debate are updated. The reason for writing the first edition was that people in the United Kingdom and the rest of Europe had been bombarded with scare stories, misinformation and half truths and found it difficult to obtain answers to their legitimate questions about genetic modification: what is it, how is it done, how does it differ from what has been done before, is it safe, how is it regulated and what implications does it have for plant breeding, agriculture and the environment? People are still asking those questions and the aim of this edition of the book, as with the first, is to provide the answers.

1 DNA, GENES, GENOMES AND PLANT BREEDING

1.1 A Brief History of Genetics

To many people the beginnings of genetics can be traced back to the publication in 1859 of Charles Darwin's book *On the Origin of Species by Means of Natural Selection*. This established the theory of evolution based on the principle of natural selection, discovered independently at the same time by Alfred Russell Wallace, and was the culmination of decades spent collecting and examining evidence.

Darwin argued that species and individuals within species compete with each other. Furthermore, individuals within a species are not all the same; they differ, or show variation. Those that are best fitted for their environment are the most likely to survive, reproduce and pass on their characteristics to the next generation. If the environment changes or a species colonises a new environment, different characteristics may be selected for, leading to change, or evolution. The diversity of life on Earth could therefore be explained by the adaptation of species and groups of individuals within species to different and changing environments, leading to the extinction of some species and the appearance of others. Species that were similar had arisen from a recent common ancestor. Most controversially, humans were similar to other apes not because God had made it so but because humans and other apes had a relatively recent common ancestor.

Darwin and Wallace were not the only scientists whose thinking was challenging the notion of life's diversity arising from supernatural creation. Georges Cuvier, in Paris, had noted that fossils in deep rock strata were less like living animals than those in shallow strata. In 1796 he published a paper on living and fossil elephants, arguing that the fossils were clear evidence of extinction. He developed his ideas further and wrote extensively on them in a series of papers, *Recherches sur les ossements fossiles de quadrupèdes*, in 1812. About the same time, studies were revealing that the anatomies of different animals were based on the same internal patterns and another French scientist, Jean-Baptiste Lamarck, had already put forward his theory that animals could transform into one another. Lamarck proposed that animals changed in response to their environment and passed these changes on to their offspring. He published his ideas in *Recherches sur l'organisation des corps vivants* in 1802, *Philosophie Zoologique* in 1809 and a series of volumes entitled *Histoire naturelle des animaux sans vertèbres* between 1815 and 1822. Although his evidence for extinction was inconsistent with creationism, Cuvier strongly opposed the notion of evolution and was a fierce critic of Lamarck. His influence probably stymied the development of evolutionary theory until the publication of *On the Origin of Species by Means of Natural Selection* half a century later.

There was a problem with the theory of evolution by natural selection and Darwin was well aware of it. At that time, the traits of parents were believed to be mixed in their offspring so that the characteristics of the offspring would always be intermediate between those of the parents. If this were true, natural selection as Darwin proposed it could not work because the process requires there to be variation within a population so that differences can be selected for. Blending traits would have the effect of reducing variation with every successive generation. Darwin even considered variations on Lamarck's theory that changes acquired during an organism's lifetime could be passed on to its offspring.

The solution of the problem was found by Gregor Mendel but, ironically, Darwin was never aware of Mendel's work. In 1857,

Mendel performed some experiments with pea plants in the garden of the Austrian monastery where he was a monk. Mendel recorded the different characteristics of the plants, such as height, seed colour, seed coat colour and pod shape, and observed that offspring sometimes, but not always, showed these same characteristics. In his first experiments, he self-pollinated short and tall plants and found that they bred true, the short having short offspring and the tall having tall offspring. However, when he crossed short and tall plants he found that all of the offspring (the F1 generation) were tall. He crossed the offspring again and the short characteristic reappeared in about a quarter of the next generation (the F2 generation).

Many years and experiments later, Mendel concluded that characteristics were passed from one generation to the next in pairs, one from each parent, and that some characteristics were dominant over others. His most famous experiments concerning the inheritance of a trait in which the seeds were wrinkled are still taught in schools today. His findings were published but ignored for decades as the work of an amateur. Later, they became known as the Mendelian laws and the foundation of modern genetics and plant breeding.

The significance of Mendel's work was that it showed that, whether or not the offspring of two parents resemble one parent or are an intermediate between the two, they inherit a single unit of inheritance from each parent. These units are reshuffled in every generation and traits can reappear, so variation is not lost. Units of inheritance subsequently became known as genes.

The next big advance came in 1902, when a British doctor, Sir Archibald Garrod, studied an inherited human disease, Alkaptonuria. Alkaptonuria sufferers excrete dark red urine because they lack an enzyme that breaks down the reddening agent, alkapton. Garrod noted that Alkaptonuria recurred in families and that parents of sufferers were often closely related to each other. In other words, it was an inherited condition. In 1902 he published 'The incidence of Alkaptonuria: a study in chemical individuality', in *The Lancet*. He continued to develop his ideas and published his most famous work, *Inborn Errors of Metabolism*, in 1909. The significance of Garrod's work was that it made the link between the inheritance of one particular gene and the activity of a single protein.

Like Mendel, Garrod was ahead of his time and his work was forgotten until the link between genes and proteins was made again in the 1930s by American geneticists George Beadle and Edward Tatum. Beadle and Tatum showed that a mutation in the fungus *Neurospora crassa* affected the synthesis of a single enzyme required to make an essential nutrient, and that the mutation was inherited through successive generations. This led to Beadle and Tatum publishing the 'one gene, one enzyme' hypothesis in 1941. The hypothesis still stands today, although it has been modified slightly to account for the fact that some proteins are made up of more than one subunit and the subunits may be encoded by different genes.

1.2 Deoxyribonucleic Acid (DNA)

With the principles of inheritance established, the search was on for the substance through which the instructions for life were passed from one generation to the next. The conclusive experiment that identified this substance is now accepted to be that conducted by Oswald Avery, Colin MacLeod and Maclyn McCarty at the Rockefeller Institute Hospital in 1944. The experiment showed that the transfer of a deoxyribonucleic acid (DNA) molecule from one strain of a bacterium, *Streptomyces pneumoniae*, to another changed its virulence. This showed that DNA was the substance that contained the information for an organism's development: the genetic material.

DNA had been discovered in 1869 by Friedrich Miescher, who worked at the Physiological Laboratory of the University of Basel and in Tübingen. All plants, animals, fungi and bacteria contain DNA and they pass a copy of their DNA to their offspring. The evolution of life on Earth was wholly dependent on the extraordinary nature of DNA and the discovery and characterisation of DNA and how it functions as the genetic material is one of the great achievements of science.

DNA consists of a backbone of units based on deoxyribose, a type of sugar, linked by phosphate groups (Figure 1.1a). In theory there is no limit to the length of the deoxyribose chain and DNA molecules can be extremely large. Attached to each deoxyribose in the chain is an organic base, of which there are four kinds: adenine, cytosine, guanine and thymine (Figure 1.1b). Any base can be present at any position and it is this that gives DNA its variability. A unit of a sugar, phosphate group and an organic base is called a nucleotide, with the different nucleotides often referred to in shorthand as A, C, G or T, according to the organic base (adenine,

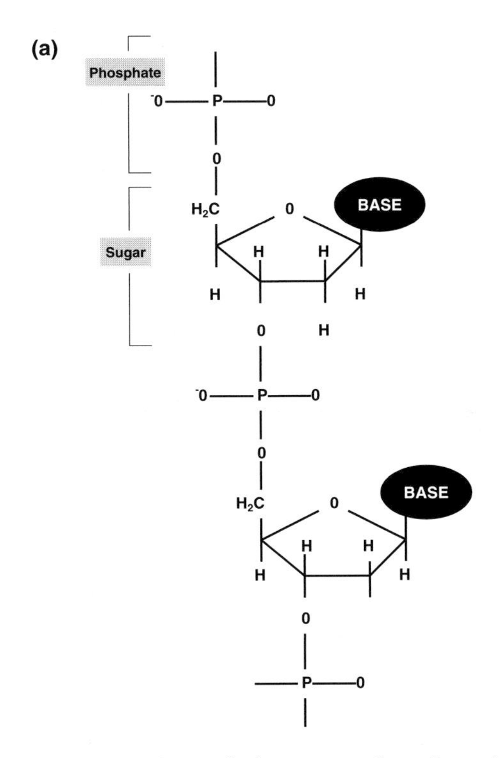

Figure 1.1 (a) Representation of the sugar-phosphate backbone in DNA. (b) Ball and stick diagram of the molecular structure of DNA. The figure shows the sugar-phosphate backbones and organic bases of the two strands and the hydrogen bonds between bases that hold the two strands together. Carbon atoms are shown black, oxygen red, phosphorous yellow, nitrogen blue and hydrogen white. Drawn by Sandra Hey.

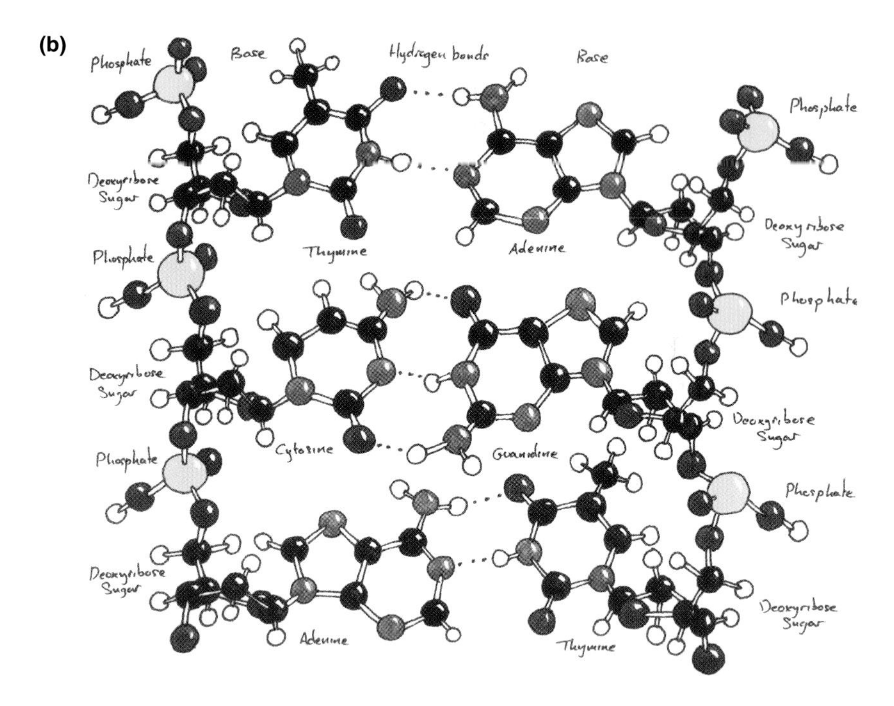

Figure 1.1 (*Continued*)

cytosine, guanine or thymine) that they contain. Information is encoded within DNA as the sequence of nucleotides in the chain. In other words, all of the instructions for life on Earth are written in a language of four letters, A, C, G and T.

The deduction of the three-dimensional structure of DNA is famously attributed to James Watson and Francis Crick, who worked together in Cambridge and published the structure in 1953. In fact, the breakthrough owed as much to the work of Maurice Wilkins and Rosalind Franklin, two scientists from a team led by William Bragg, who were pioneering the techniques of X-ray crystallography in the study of large molecules in Cambridge at that time. Watson and Crick were inspired by the work of Linus Pauling, who had discovered that the molecules of some proteins have helical shapes. They designed several models of DNA and finally came up with the correct structure after analysing Franklin's X-ray photographs. Watson, Crick and Wilkins jointly received the Nobel Prize for Physiology or Medicine in 1962. Tragically, Rosalind Franklin died of cancer in 1958 and could not be honoured because the Nobel Prize is not awarded posthumously.

The structure of DNA, as deduced by Watson and Crick, consists of two deoxyribose chains running in opposite directions coiled around each other to form a beautiful structure called a double helix (Figure 1.2). The bases attached to each deoxyribose are on the inside of the helix at right angles to the helix axis and the structure is stabilised by hydrogen bonds between adjacent bases from each chain (Figures 1.1b and 1.2). Because of the dimensions of the helix and the structures of the bases, adenine on one chain must always be faced with thymine on the other, while guanine is always paired with cytosine. These pairs, A and T or C and G, are often referred to as base pairs, and the size or length of a DNA molecule is often given in base pairs (bp).

This specific pairing of the bases is important because it underlies the process by which DNA can be replicated as an exact copy.

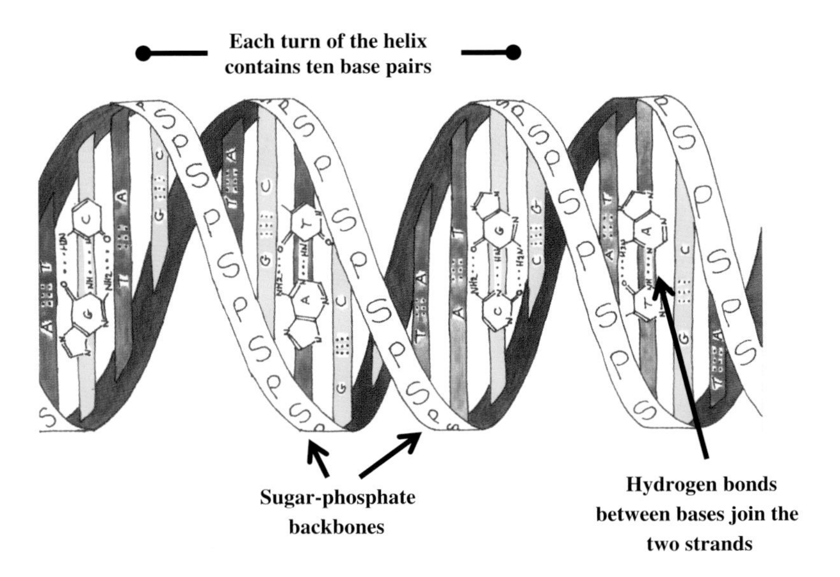

Figure 1.2 Structure of a DNA double helix. S = sugar, P = phosphate. The organic bases, adenine, cytosine, guanine and thymine, are labelled A, C, G and T. Drawn by Sandra Hey.

If double-stranded DNA is unravelled to form two single strands, each strand can act as a template for the synthesis of a complementary chain and two replicas of the original double-stranded molecule are created.

1.3 Genes

We can now define genes, the units of heredity, as functional units within a DNA molecule. A gene will contain the information not only for the structure of a protein, but also for when and where in an organism the gene is active. This information is encoded in the sequence of base pairs within the region of the DNA molecule that makes up the gene. Genes can comprise over a million base pairs but are usually much smaller, averaging 3,000 base pairs.

The part of the gene containing the information for the primary structure of the protein encoded by the gene itself is called 'the coding region' (Figure 1.3). The sequence of amino acids in the protein is determined by the sequence of base pairs in the coding region of the gene, each amino acid in the protein being represented by a triplet of base pairs in the DNA sequence. Adjacent to (or 'up-' and 'downstream' of) the coding sequence are the regulatory regions of the gene that determine when and where the gene

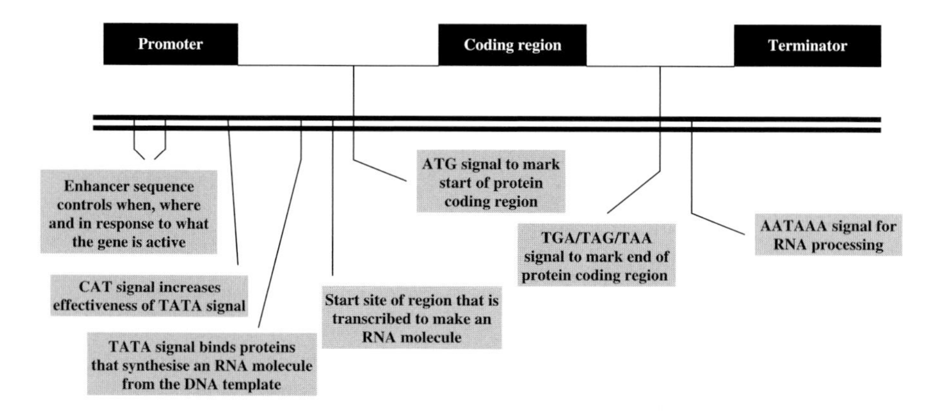

Figure 1.3 Schematic diagram of the structure of a gene.

is active (or expressed). This regulatory information is usually, but not always, 'upstream' of the coding region in what is called the gene promoter. The region 'downstream' of the coding region is called the gene terminator.

1.4 Gene Expression

The term *gene expression* refers to the process by which the protein that is encoded by the gene is produced. It comprises the following sequence of events:

1. The DNA molecule is used as a template for the synthesis of a related, single-stranded molecule called ribonucleic acid (RNA). RNA, like DNA, comprises organic bases on a sugar-phosphate backbone. The sequence of bases on the newly synthesised RNA molecule is determined by the sequence of bases on the DNA template, hence the encoded information is passed from the DNA to the RNA molecule. This process is called transcription.
2. The RNA molecule is processed and transported to protein complexes called ribosomes where protein synthesis occurs.
3. A protein is synthesised, the amino acid sequence of which is specified by the RNA molecule. This process is called translation.

At the end of this process, information encoded in the DNA molecule has been transferred through an RNA molecule to the protein production machinery and used to make a protein.

Genes that are active (i.e. being expressed) throughout an organism all of the time are referred to as constitutive or housekeeping genes. Most genes are not constitutive, but are subject to various kinds of regulation. Some are expressed only in certain organs, tissues or cell types. Others are expressed during specific developmental stages of an organism, or are expressed in response to a stimulus. In the case of plants, gene expression responds to a host of stimuli, including light, temperature, freezing, grazing, disease, shading and nutritional status. DNA is not just the blue-print for an organism's development; it is active throughout an organism's life and plays a central role in an organism's interaction with its environment.

1.5 Genomes

The genome is defined as an organism's complete complement of DNA. Genomes can be very large but it is now possible to determine the sequence of base pairs for entire genomes from even complex animals, plants and fungi. The entire yeast genome base pair sequence was determined in 1996 and a draft of the human genome sequence was completed in 2000. The base pair sequences of entire genomes have also been determined for Arabidopsis (*Arabidopsis thaliana*), a plant also known as thale cress that is used as a model in plant genetics because of its relatively small genome, Brachypodium (*Brachypodium distachyon*), a species of grass used as a model for cereals, also because of its relatively small genome, and rice (*Oryza sativa*). There are emerging genome sequences for several plant species, including soybean (*Glycine max*), lotus (*Lotus japonicus*), poplar (*Populus trichocarpa* spp.), sorghum (*Sorghum bicolor*), wine grape (*Vitis vinifera*), maize (*Zea mays*), potato (*Solanum tuberosum*) and wheat (*Triticum aestivum*). The technology for determining the sequence of base pairs in DNA is still developing rapidly, and so-called 'next generation sequencing' (NGS) technologies are enabling DNA sequence information to be generated at a throughput that was not dreamt of even ten years ago. This holds out the prospect in the next few years of being able to determine the base pair sequence for individual genomes, with applications in human and animal disease research and diagnosis as well as plant breeding. There may still be limitations for very large genomes arising from the computing capacity required to handle and analyse the data. Indeed, that is currently the case, for example, for wheat; although sequence data covers the entire genome it is still not possible to assemble it all, making the data difficult to access and use.

The human genome contains approximately three billion base pairs, organised into 23 chromosomes ranging from 50–250 million base pairs long. The rice genome contains 466 million base pairs, while that of Arabidopsis contains approximately 126 million base pairs and that of wheat a daunting 16 billion base pairs. The human genome, however, contains only 30–40,000 genes, while that of rice contains 45–56,000 genes, with wheat containing a similar number (one could argue that rice and wheat are more complex and highly evolved organisms than humans). The reason for the discrepancy between genome size and gene number is that much of the genome does not contain genes. In fact, genes represent only 2% of the human genome. The rest of the DNA is sometimes referred to as 'junk DNA' and the amount of it varies greatly between different species. Much of it is highly repetitive in its base pair sequence and it may promote rearrangement of the DNA molecules, driving the formation of new genes and therefore variation and evolution.

1.6 Genetic Change

Genetic change is a natural and desirable process. It results in variation in shape, form and behaviour of the individuals within a species, allowing for evolution and adaptation. It is crucial for the survival of any species and the evolution of new species in response to environmental change.

The process of genetic change in nature is called evolution and when driven by natural selection it results in adaptive improvement. In the absence of selection it leads to diversity and variation, making it more likely that some individuals will survive if the environment changes.

1.7 Plant Breeding

Natural selection is not the only driver for genetic change at work in the world today. Man has discovered that rapid genetic change can be induced by the artificial selection of individuals for breeding. This has been used to produce the Chihuahua, the greyhound and the bull terrier from the wolf, thoroughbred racehorses, the shire horse and the Shetland pony, and every

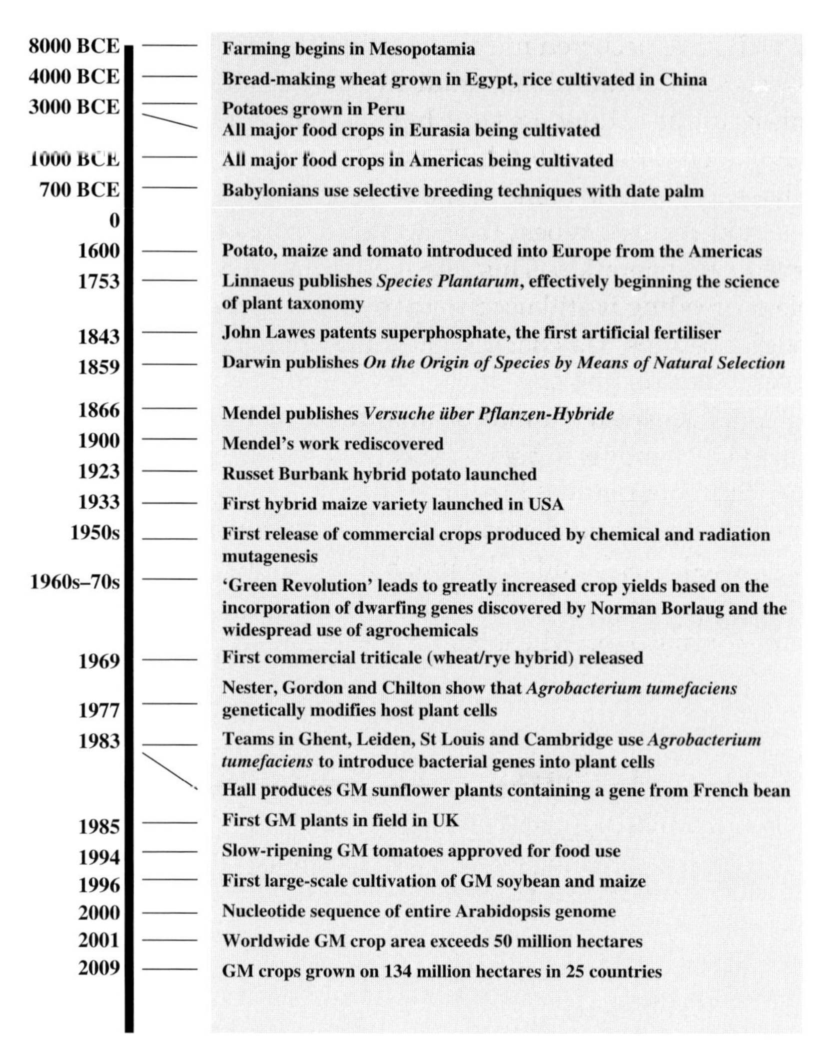

Figure 1.4 Timeline: 10,000 years of farming and plant breeding.

modern farm animal. It has also been applied for millennia to crop improvement (Figure 1.4).

It is probable that crop improvement has been practised since humankind first started to plant and harvest crops, rather than forage for food from wild plants. This is thought to have become widespread about 10,000 years ago. At first, such improvement may well have occurred unconsciously, through the selection of the most vigorous individuals from highly variable populations for planting in the following year, but then became more systematic. Types of a crop plant with different characteristics would be grown in adjacent plots and some of the seed produced would result from crossing of the two types. Farmers would then select the best seed for the next generation. This relatively primitive but effective form of plant breeding is still used in many parts of the world today and through the ages has changed crop plants greatly from their wild ancestors and relatives.

This is illustrated quite dramatically in Figure 1.5, in which grain from a modern bread-making wheat variety is compared with grain from two wild relatives. Grain that was buried in tombs in Ancient Egypt and discovered in modern times when the tombs were opened was found to be much more similar to modern wheat grain than to grain of wild wheat. So, even by that time, 4,000 or more years ago, wheat had been changed greatly by simple

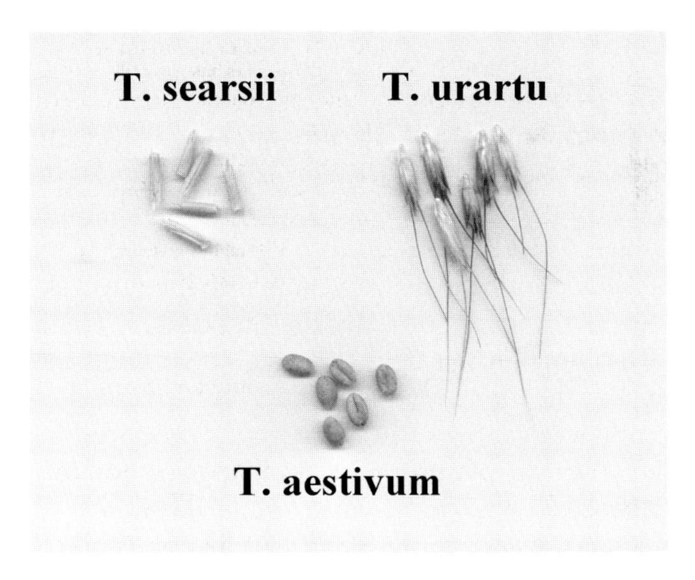

Figure 1.5 Grain of bread wheat (*Triticum aestivum*) compared with that of two wild relatives, *Triticum searsii* and *Triticum urartu*. Bread wheat is a product of thousands of years of selective breeding.

selection. Wheat cultivation and improvement enabled the production of food in large quantities and it is doubtful whether modern western civilisation would have developed without it; a similar argument could be made for the dependence of eastern civilisation on rice. The development of modern wheat is a good example of how the genetics of crop plants has been manipulated by farmers and breeders who for millennia knew nothing about the molecular basis of what they were doing.

Wheat actually comprises many different species, some with one, two or even three genomes (diploids, tetraploids and hexaploids, respectively). The tetraploids, which include cultivated pasta wheats, arose through hybridisation and genome duplication between ancestral diploid species. Bread-making wheat is a hexaploid that arose through hybridisation between a tetraploid and a third diploid species. There are no wild hexaploid wheat species: bread-making wheat appeared within cultivation in south-west Asia approximately 10,000 years ago and its use spread westwards into Europe.

Another good example of the remarkable changes that have been introduced into crop plants by simple selection over many generations is the cabbage family of vegetables, which includes kale, cabbage, cauliflower, broccoli and Brussels sprouts (Figure 1.6). The wild relative of the cabbage family grows in the Mediterranean region of Europe, and it was first domesticated approximately 7,000 years ago. Through selective breeding, the crop plants became larger and leafier, until a plant very similar to modern kale was produced in the fifth century BCE. By the first century CE, a different variation was being grown alongside kale. It had a cluster of tender young leaves at the top of the plant and is known today as cabbage.

In the fifteenth century, cauliflower was produced in southern Europe by selecting plants with large, tender, edible flowering heads and broccoli was produced in similar fashion in Italy about a century later. The last variation to appear was Brussels sprouts, which were bred in Belgium in the eighteenth century, with large buds along the stem. All of these very different vegetables are actually the same species, *Brassica oleracea*.

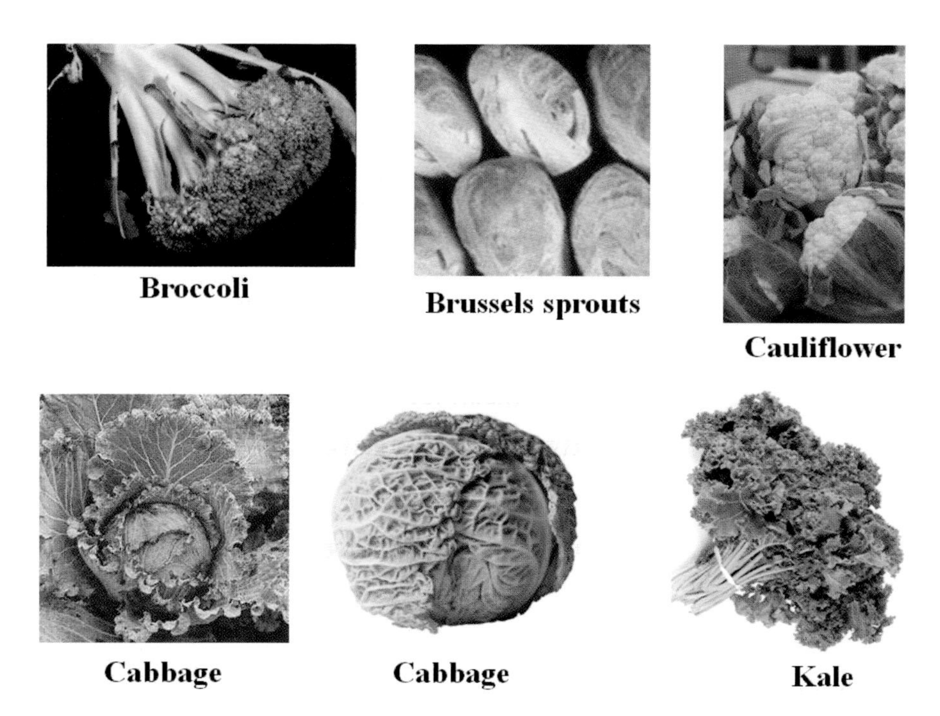

Figure 1.6 The cabbage family, all forms of the same species: *Brassica oleracea*.

1.8 Modern Plant Breeding

True scientific breeding dates from about 1900 (Figure 1.4) when the rediscovery of the work of Gregor Mendel on the inheritance of characteristics in pea provided a sound theoretical basis. As stated above, plants contain many genes, estimated at about 26,000 in Arabidopsis and 45–56,000 in rice. These genes may exist in different forms (called alleles) in individuals of the same species, in the same way, for example, that alleles of the gene that imparts eye colour exist in humans. Crossing of individuals with contrasting characteristics results in a population of individuals with different combinations of alleles from the two parents. At least some of these combinations may result in improved performance relative to either parent, and the identification and further improvement of these is the task of the plant breeder. Since the breeder is literally dealing with thousands of genes, the task is formidable.

Nevertheless, plant breeders have been very effective and it is just as well that they have. At the end of the eighteenth century, Reverend Thomas Malthus predicted in his 'Essay on the principle of population', which he published anonymously, that food supply could not keep up with exponentially rising population growth. At the time, world population was approximately one billion. In 1999, the world population reached six billion and it will pass seven billion in 2011, and yet, with the exception of sub-Saharan Africa, that population has generally not been critically short of food. Food demand has been met through dramatic increases in crop yield. As an example, the yield per hectare of wheat grown in the UK from approximately 1200 to the 1970s is shown in Figure 1.7. It increased approximately ten-fold, with more than half that increase coming after 1900. Wheat yield today in the UK regularly exceeds 10 tonnes per hectare. Of course, plant breeding has not been the only

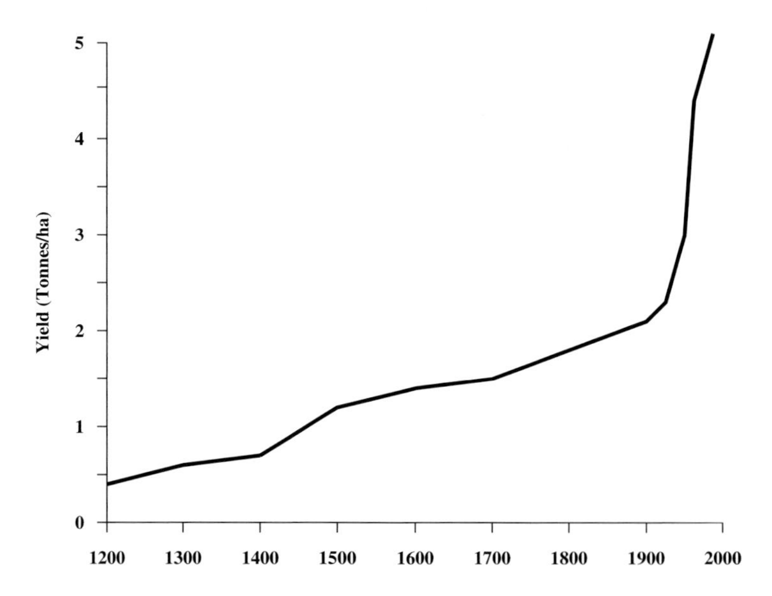

Figure 1.7 United Kingdom wheat yield, 1200 to 1970s (data from United States Department of Agriculture). Wheat yield in the UK now regularly exceeds 10 tonnes per hectare.

contributor to this increase. Mechanisation, the development and use of nitrogen fertilisers, herbicides and pesticides and other improvements in farm practice have played a part. Indeed, it was the combination of mechanisation, the widespread use of agro-chemicals and the incorporation of dwarfing genes into cereals that led to the 'Green Revolution' of the 1960s and 1970s. The dwarfing genes concerned actually affected the synthesis of a plant hormone, gibberellin, although that was not known at the time. Their incorporation reduced the amount of resources that crop plants put into their inedible parts and at the same time made crops less susceptible to lodging (falling over in damp and/or windy conditions).

Plant breeding has also been successful in improving food safety. Contrary to the popular notion that natural is 'good' and man-made is 'bad' when it comes to food, plants did not evolve to be eaten. In fact, with the exception of fruit, the opposite is true, and since plants are unable to flee a large animal or brush off an insect and do not have an immune system like that of animals to fight off micro-organisms, they have evolved a sophisticated armoury of chemical weapons with which to defend themselves. Many of these chemicals are potentially toxic or allergenic to humans and they are most abundant in seeds and tubers, whose rich reserves of proteins, starch and oil are particularly attractive to herbivores, pests and pathogens. Well-known examples are glycoalkaloids in potatoes, cyanogenic glycosides in linseed, proteinase inhibitors in soybean and other legume seeds and glucosinolates in *Brassica* oilseeds.

Plant breeders have made great strides in reducing the levels of these chemicals in crop plants, although some crops have only been regarded as fit for human consumption surprisingly recently. Oilseed rape, for example, was first grown in the UK during World War II to provide oil for industrial uses and some varieties are still grown for that purpose. Its oil was regarded as unfit for human consumption because it contained high levels of erucic acid and glucosinolates. Erucic acid (see Chapter 3) is a fatty acid that has been linked with the formation of fatty deposits in heart muscle and consequent muscle damage. Glucosinolates have a bitter/hot flavour and are poisonous, although their toxicity is complicated by the fact that they break down in the body to a variety of different compounds, with different glucosinolates producing different breakdown products. It has been suggested that they are responsible for oilseed rape having relatively few specialised pests. Glucosinolates are present at even higher levels in the meal that remains after oil has been extracted, with implications for the use of the meal in animal feed. After World War II, breeders gradually reduced the levels of erucic acid and glucosinolates, but the oil was only passed for human consumption and the meal for animal feed in the 1980s.

1.9 Wide and Forced Crossing and Embryo Rescue

There is a major limitation to the success of plant breeding, which is the extent of variation in the parental lines. Plant breeders cannot select for variation which is not present in their breeding population. They have therefore looked further afield, to exotic varieties and wild relatives of crop species, for genes to confer important characters. This 'wide crossing' would not occur in nature and usually requires 'rescue' of the embryo to prevent abortion. This entails surface-sterilising developing seeds and dissecting them under a microscope to remove the embryos. The embryos are then cultured in a nutrient medium until they germinate and develop into seedlings, which are then transferred to soil.

Forced crosses of this sort may be carried out to transfer genes for single characters (disease resistance, for example). However, tens of thousands of other genes will inevitably be brought into the breeding programme alongside the desired gene. Most of these can be eliminated by crossing the hybrid plant repeatedly with the crop parent (back-crossing). Nevertheless, many genes of unknown function will remain at the end of this process and if they have undesirable effects on the performance of the crop the programme will fail.

Forced crosses have even been made between different plant species. The best-known example of this is triticale, a hybrid between wheat and rye. The first deliberately made hybrid between wheat and rye was reported by A.S. Wilson in Scotland in 1875. However, these plants were sterile because the chromosomes from wheat and rye will not form pairs during meiosis. This can be overcome by inducing chromosome doubling by chemical treatment, usually with colchicine. The cross is usually made between durum wheat, a tetraploid, and rye to produce a hexaploid triticale, although it is also possible to cross hexaploid wheat with rye to produce an octoploid triticale.

The name triticale was used first in 1935 by Tschermak, but it was not until 1969, after considerable improvement through breeding, that the first commercial varieties of triticale were released. Triticale is now grown on more than 2.4 million hectares worldwide, producing more than six million tonnes of grain per year. Its advantage is that it possesses the yield potential of wheat and the hardiness (tolerance of acid soils, damp conditions and extreme temperatures) of rye. However, it does not match the bread-making quality of wheat and up to now has been used mostly for animal feed. It may have potential as a raw material for bioethanol production.

1.10 Radiation and Chemical Mutagenesis

Another way of increasing the variation within a breeding population is to introduce mutations artificially. This is usually done with seeds and involves treatment with neutrons, gamma rays, X-rays, UV radiation or a chemical (a mutagen). All of these treatments damage DNA. If there is too much damage the seed will die, but minor damage can be repaired, resulting in changes in the DNA sequence. Sometimes even small changes of this sort are lethal, but occasionally changes are made to a gene that leave the seed viable but alter the characteristics of the plant. The process is entirely random and mutagenesis programmes usually involve very large populations of at least 10,000 individuals.

The first attempts to produce plant mutants were made in the 1920s and the first commercial varieties arising from mutation breeding programmes became available in the 1950s. The technique was particularly fashionable in the 1960s and 1970s but continues to be used today. One of the earliest cultivars produced by mutagenesis was the oilseed rape cultivar Regina II, which was released in Canada in 1953. Mutagenesis played an important role in the improvement of oil quality of oilseed rape and flax, and in durum wheat breeding in Italy. There are believed to be hundreds of rice varieties that incorporate artificially induced mutations and most North American white bean varieties incorporate a mutation that was first induced by X-ray treatment.

The best-known example of an irradiation mutant grown in the UK is the barley variety Golden Promise. This was the most successful UK malting barley variety in the latter part of the last century and although it is not grown widely now it has been incorporated into many barley breeding programmes.

1.11 The Advent of Genetic Modification

Crossing different varieties and species involves the mixing of tens of thousands of genes, sometimes with unpredictable results, while mutagenesis is, of course, a random process and it is impossible to count, let alone characterise, all of the effects of a mutagenesis programme. Radiation and chemical mutants are likely to carry a baggage of uncharacterised genetic changes with unknown effects. Nevertheless, these techniques were developed and applied to plant breeding at a time when food production was a much higher priority than food safety and the risks were not a matter of public debate. They have now been around for so long that they are considered to be part of 'traditional' plant breeding and crops modified by wide crossing and mutation breeding are accepted readily.

Even with the possibility of crossing crop plants with exotic relatives and related species and introducing mutations artificially, the variation available to plant breeders remained limited. Both techniques also had the disadvantage of introducing unwanted as well as desirable genetic changes. In the late 1970s, a new technique became available, that of inserting specific genes into the genome of a plant artificially. It meant that technically there was no limit to the source of new genes and it enabled plant breeders to bring specific genes into breeding programmes without unwanted genetic baggage. This new technique was first called genetic engineering and subsequently genetic modification.

2 THE TECHNIQUES OF PLANT GENETIC MODIFICATION

Any new crop variety has been changed genetically, using the methods described in the previous chapter. However, the term genetically modified (GM) is a relatively new expression that describes a plant that contains a gene or genes that have been introduced artificially. Such plants are also described as being transgenic or having been transformed. The term genetically engineered (GE) can be used instead of genetically modified. Plant genetic modification became possible in the late 1970s as a result of the development of techniques for manipulating DNA in the laboratory and introducing it into the DNA of a plant.

2.1 A Brief History of the Development of Recombinant DNA Technology

In the quarter-century after the description of the structure of DNA by Watson and Crick in 1953, huge strides were made in the study of DNA and the enzymes present in cells that can work on it (Figure 2.1). In 1955, Arthur Kornberg at Stanford University isolated DNA polymerase, an enzyme that synthesises DNA. He received the Nobel Prize for Medicine in 1959. In 1966, Bernard Weiss and Charles Richardson working at Johns Hopkins University isolated DNA ligase, an enzyme that 'glues' two ends of DNA together. In 1970, Hamilton Smith, also at Johns Hopkins

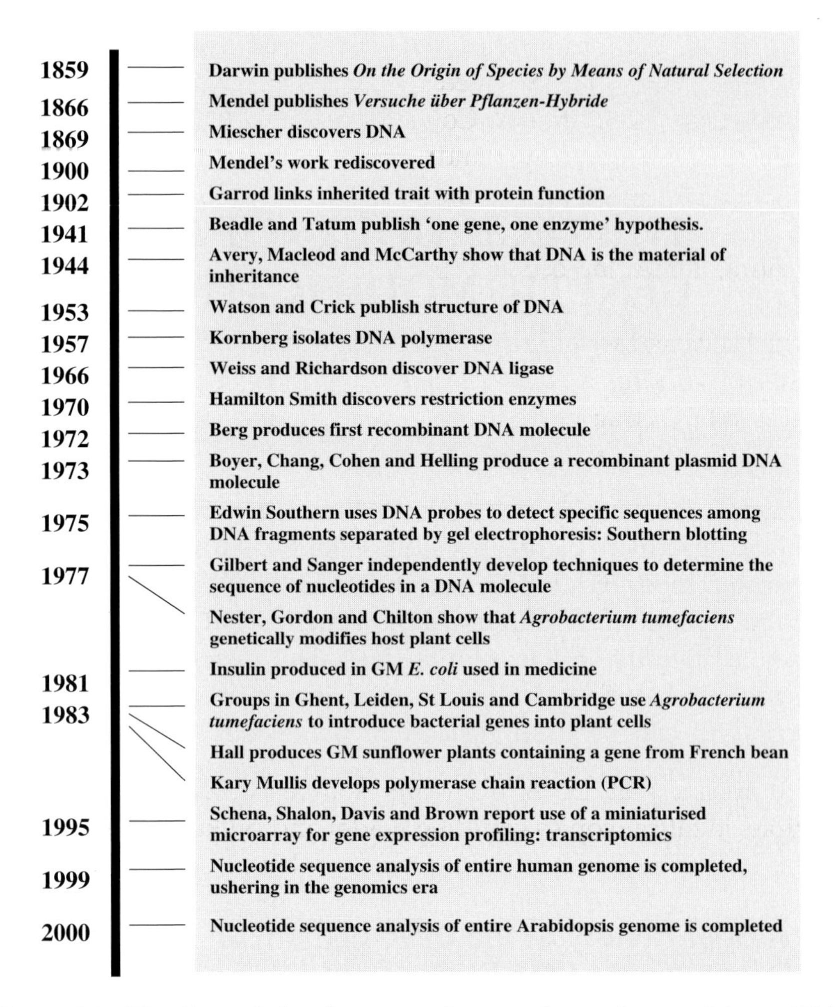

Figure 2.1 Timeline of developments in genetics and recombinant DNA technology.

University, described the first characterisation of a restriction endonuclease (now usually called a restriction enzyme), an enzyme that had the ability to recognise specific short sequences of base pairs in a DNA molecule and cut the molecule at that point. Smith shared the Nobel Prize for Medicine in 1978.

The rapid progress continued. In 1972, Paul Berg at Stanford reported that he had constructed a DNA molecule by cutting viral and bacterial DNA molecules with restriction enzymes and then recombining them. He received the Nobel Prize for Chemistry in 1980. Then in 1973, Stanley Cohen and Annie Chang of Stanford together with Herbert Boyer and Robert Helling of the University of California, San Francisco, demonstrated that DNA that had been cut with a restriction enzyme could be recombined with small, self-replicating DNA molecules from bacteria, called plasmids. The product, a recombinant plasmid, could then be reintroduced into bacterial cells and would replicate. In 1975, a British scientist, Edwin Southern, working at the University of Edinburgh, reported the transfer of DNA fragments that had been separated by electrophoresis to a membrane and the detection of specific DNA fragments by hybridisation with DNA probes. This technique became known as Southern blotting. Then, in 1977, Walter Gilbert at Harvard and Fred Sanger in Cambridge separately developed methods for determining the sequence of nucleotides in a DNA molecule. Sanger and Gilbert shared the Nobel Prize for Chemistry with Berg in 1980.

Scientists now had the tools for cutting DNA molecules at specific points and gluing them back together in different combinations to make new molecules. This was known as recombinant DNA technology. It meant that DNA from any source could be cloned (i.e. multiple, identical copies could be made) and bulked up in bacteria for analysis. The bacterium of choice for this purpose was and is *Escherichia coli*. This is a human gut bacterium, although the strains used in the laboratory have been modified so that they are not pathogenic. The cutting and splicing of plant, animal and fungal genes into plasmids and their propagation in *Escherichia coli* underpinned the molecular analysis of gene structure and function. It also led to the first commercial use of recombinant DNA technology in the pharmaceutical industry, recombinant human insulin produced in *Escherichia coli* being approved by the Food and Drug Administration of the USA in 1981.

The technology continued to develop. In 1983, Kary Mullis developed the polymerase chain reaction, which uses a DNA polymerase from a thermophilic bacterium. This polymerase survives boiling and has an optimum temperature for activity of 72°C, allowing for repeated heating and cooling of the reaction. Specific DNA fragments are amplified from extremely small amounts of template DNA to produce sufficient amounts for cloning or direct analysis. Then, in 1995, miniature microarrays were developed that allow the expression levels of tens of thousands of genes in an organism, organ or cell type to be measured in a single experiment. Microarrays consist of thousands of microscopic spots of DNA oligonucleotides (short DNA fragments), each matching a specific gene nucleotide sequence. These are probed with total RNA populations from the tissue or organ being investigated. The development of microarrays was dependent on the increasing genomic data that was coming into the public domain at that time. This culminated with the publication of the entire human genome nucleotide sequence in 1999 and, not far behind, that of the model plant, Arabidopsis (*Arabidopsis thaliana*), in 2000.

2.2 *Agrobacterium tumefaciens*

Escherichia coli is the bacterium of choice for general DNA cloning and manipulation. Another bacterial species that contains a plasmid is the common soil bacterium, *Agrobacterium tumefaciens*, a bacterium that infects wounded plant tissue and causes the disease known as crown gall. In 1977, Eugene Nester, Milton Gordon and Mary-Dell Chilton showed that genes from a plasmid carried by *Agrobacterium tumefaciens* were inserted into the DNA of host plant cells as a natural part of the infection process. Effectively, the bacterium could perform the genetic modification of host plant cells. The plasmid was called the tumour-inducing or Ti plasmid. This observation led to *Agrobacterium tumefaciens* becoming one of the most reliable and widely used means of transferring foreign DNA (DNA from a different organism) into plants.

During infection, the host plant cells release phenolic compounds. These compounds interact with proteins encoded by virulence (*VIR*) genes carried by the Ti plasmid and induce the bacteria to bind to plant cell walls. Other *VIR* genes on the Ti plasmid then become active and they cause a single strand of DNA to be nicked out of a region of the Ti plasmid. This T-DNA (transfer DNA) is protected by specialised proteins and is transferred to the adjacent plant cell where it integrates into the plant DNA.

Once integrated into the plant DNA, genes present in the T-DNA become active. These genes encode proteins that perturb the normal hormone balance of the cell. The cell begins to grow and divide to form a tumour-like growth called the crown gall. The cells of the crown gall are not differentiated, in other words they do not develop into the specialised cells of a normal plant. They can be removed from the plant and cultured as long as they are supplied with light and nutrients and are protected from fungal and bacterial infection. A clump of these undifferentiated cells is called a callus.

All of the cells of a crown gall or a callus contain the T-DNA that was inserted into the original host cell. Indeed, the T-DNA will be inherited stably through callus culture and through generations of GM plants that are produced from the callus. The T-DNA contains another set of genes that induce the host cell to make and secrete unusual sugar and amino acid derivatives that are called opines, on which the *Agrobacterium* feeds. There are several types of opine, including nopaline and octopine, and different types are produced after infection with different strains of the bacterium. The genes that enable the bacterium to feed on the opines are also carried by the Ti plasmid.

Agrobacterium tumefaciens has a close relative, *Agrobacterium rhizogenes*, that causes hairy root disease. *Agrobacterium rhizogenes* uses a similar strategy to *Agrobacterium tumefaciens*, and carries an Ri plasmid that functions in a similar way to the Ti plasmid but induces a different response in the host plant cell. Together they are nature's genetic engineers and were using plant genetic modification millions of years before humans 'invented' it.

2.3 Use of *Agrobacterium tumefaciens* in Plant Genetic Modification

The natural mechanism of plant genetic modification by *Agrobacterium tumefaciens* can be used to introduce any gene into a plant cell. This is because the only parts of the T-DNA that are required for the transfer process are short regions of 25 base pairs at each end, or border. Anything between these border regions will be transferred into the DNA of the host plant cell. Callus formation can be induced in the laboratory by infecting leaf pieces, stem sections, tuber discs or other 'explants' with *Agrobacterium tumefaciens*. In 1983, groups led by Jeff Schell and Marc Van Montagu in Ghent, Rob Schilperoort in Leiden, Mary-Dell Chilton and Michael Bevan in St Louis and Cambridge and Robert Fraley, Stephen Rogers and Robert Horsch, also in St Louis, showed that bacterial antibiotic resistance genes could be inserted into the T-DNA carried on a Ti plasmid and transferred into plant cells. In the same year, Tim Hall and colleagues, working at the Agrigenetics Advanced Research Laboratory in Wisconsin and the University of Wisconsin, used this method to produce a sunflower plant carrying a seed protein gene from French bean. Not only was the gene present in every cell of the plant, but it was inherited stably and was active. This was the first report of what would later be called a genetically modified or GM plant.

Michael Bevan in Cambridge then developed so-called binary vectors, plasmids that would replicate in both *Escherichia coli*, in which they could be manipulated and bulked up, and *Agrobacterium tumefaciens*. Binary vectors contain the left and right T-DNA borders but none of the genes present in 'wild-type' T-DNA. They are unable to induce transfer of the T-DNA into a plant cell on their own because they lack the *VIR* genes that are required to do so. However, when present in *Agrobacterium tumefaciens* together with another plasmid containing the *VIR* genes, the region of DNA between the T-DNA borders is transferred, carrying any genes that have been placed there.

The explants are transferred to sterile dishes containing solid plant growth medium and cells that have received the T-DNA divide to form calli (Figure 2.2a). Calli are then transferred onto medium that contains a plant hormone that induces them to form shoots (Figure 2.2b). Once a shoot with a stem has formed, it is transferred to a medium that does not contain the shoot-inducing

Figure 2.2 ***Agrobacterium tumefaciens*-mediated transformation of potato.**
a. Potato tuber discs that have been infected with *Agrobacterium tumefaciens*, showing the formation of clumps of undifferentiated cells called callus. Cells within the callus contain genes that have been inserted into the potato genome by the bacterium.
b. Shoots induced from callus forming on leaf pieces by the application of a plant hormone.
c. A complete GM potato plantlet.
d. Once the plantlets are large enough they can be transferred to soil in pots and grown in containment in a greenhouse.

hormone. Hormones produced by the shoot itself then induce root formation and a complete plantlet is formed (Figure 2.2c). Up to this point the process has to be done in sterile conditions to prevent bacterial or fungal infection. Once shoot and root are fully formed, the plantlet can be transferred to soil and treated like any other plant (Figure 2.2d).

Plants that have been altered genetically in this way are referred to as transformed, transgenic or genetically engineered, as well as genetically modified. The term transgenic is favoured by scientists but genetically modified has been adopted most widely by non-specialists. All plant breeding, of course, involves the alteration (or, if you like, modification) of plant genes, whether it is through the selection of a naturally occurring mutant, the crossing of different varieties or even related species or the artificial induction of random mutations through chemical or radiation mutagenesis. Nevertheless, the term genetically modified is now used specifically to describe plants produced by the artificial insertion of a single gene or small group of genes into its DNA. Genetic modification has been an extremely valuable tool in plant genetic research as well as in crop plant breeding.

2.4 Transformation of Protoplasts

Protoplasts are plant cells without a cell wall. The cells are usually derived from leaf tissue and are incubated with enzymes (cellulases, pectinases and hemicellulases) that digest away the cell wall. *Agrobacterium tumefaciens* will infect and transform protoplasts; the protoplasts can then be cultured and induced to form calli by the application of plant hormones (auxins and cytokinins). GM plants can be regenerated from the calli as described for *Agrobacterium tumefaciens*-mediated transformation of explant material.

Protoplasts can also be induced to take up DNA directly. This process is called direct gene transfer or DNA-mediated gene transfer. There are two widely used methods for achieving this. The first involves treatment with polyethylene glycol (PEG) or a similar polyvalent cation. The exact way in which PEG works is not known, but it is believed to act by causing the DNA to come out of solution and by eliminating charge repulsion. The second method is called electroporation and involves subjecting the protoplasts to a high voltage pulse of electricity. This causes the formation of pores in the plant cell membrane and, although these must be repaired very rapidly to prevent the protoplasts from dying, it briefly allows DNA to be taken up by the protoplasts.

Direct gene transfer into protoplasts is most commonly used as a research tool. A gene of interest is introduced into protoplasts so that its activity and function can be studied. The gene does not integrate into the protoplasts' own DNA and is eventually broken down, so the protoplasts are only temporarily, or transiently, rather than stably, transformed. However, in a small proportion of the protoplasts, the introduced DNA will integrate into the host DNA and the protoplast will be stably transformed. The protoplast can then be induced to form callus and a GM plant can be regenerated from it.

There are two major drawbacks to the production of GM plants by this method. Firstly, it is not possible to induce protoplasts of all plant species to form calli and regenerate a whole plant. Secondly, the introduced DNA is often rearranged and does not function as expected.

2.5 Particle Gun

Flowering plants can be subdivided into two classes, monocotyledonous and dicotyledonous, depending on their embryo structure. In the wild, *Agrobacterium tumefaciens* only infects dicotyledonous plants and although progress has since been made in adapting its use for the genetic modification of monocotyledonous plants in the laboratory, and it is now the method of choice for some monocotyledonous species, its use was limited at first to dicotyledonous plants. This was important because cereals, including major crop species such as wheat, maize and rice, belong to the monocotyledonous class and these species are also not amenable to the regeneration of whole plants from protoplasts.

Transformation of cereals eventually became possible with the invention of the particle bombardment method. In this method, plant cells are bombarded with tiny particles coated with DNA. In some cases, the wall of the cell is penetrated without the cell being killed. Some of the DNA is washed off the particles and becomes integrated into the plant genome. This is carried out in a particle gun, the first of which used a small explosive charge to bombard the plant cells with tungsten particles. Subsequently, a variety of devices were developed but one using a burst of pressurised helium gas in place of the explosive charge has been the most successful and gold particles are now used instead of tungsten. There are now hand-held versions of this device on the market, allowing DNA to be delivered to cells within intact plants. This has become an extremely useful tool for studying the transient activity and function of genes that are introduced into plant cells and remain there for a short period of time but do not integrate into the host plant DNA.

Particle bombardment, which has acquired the name of biolistics, has been particularly successful in the production of genetically modified cereals, including maize, wheat, barley, rice, rye and oat, and the first commercial GM cereal crops (Chapter 3) were produced using this method. The plant tissues that are used are either isolated explants (Figure 2.3a) that are bombarded, induced to become embryogenic and regenerated, or embryogenic cell cultures. Whole plants are then regenerated from single embryonic cells that have taken up the DNA (Figures 2.3b–d).

2.6 Other Direct Gene Transfer Methods

The technique of electroporation can be applied to intact cells in tissue pieces or in suspension, as well as to protoplasts, but has only been shown to work efficiently in a few species. The other direct gene transfer method to have been developed is silicon carbide fibre vortexing. Plant cells are suspended in a medium containing DNA and microscopic silicon carbide fibres. The suspension is vortexed and the fibres penetrate the plant cell walls, allowing the DNA to enter.

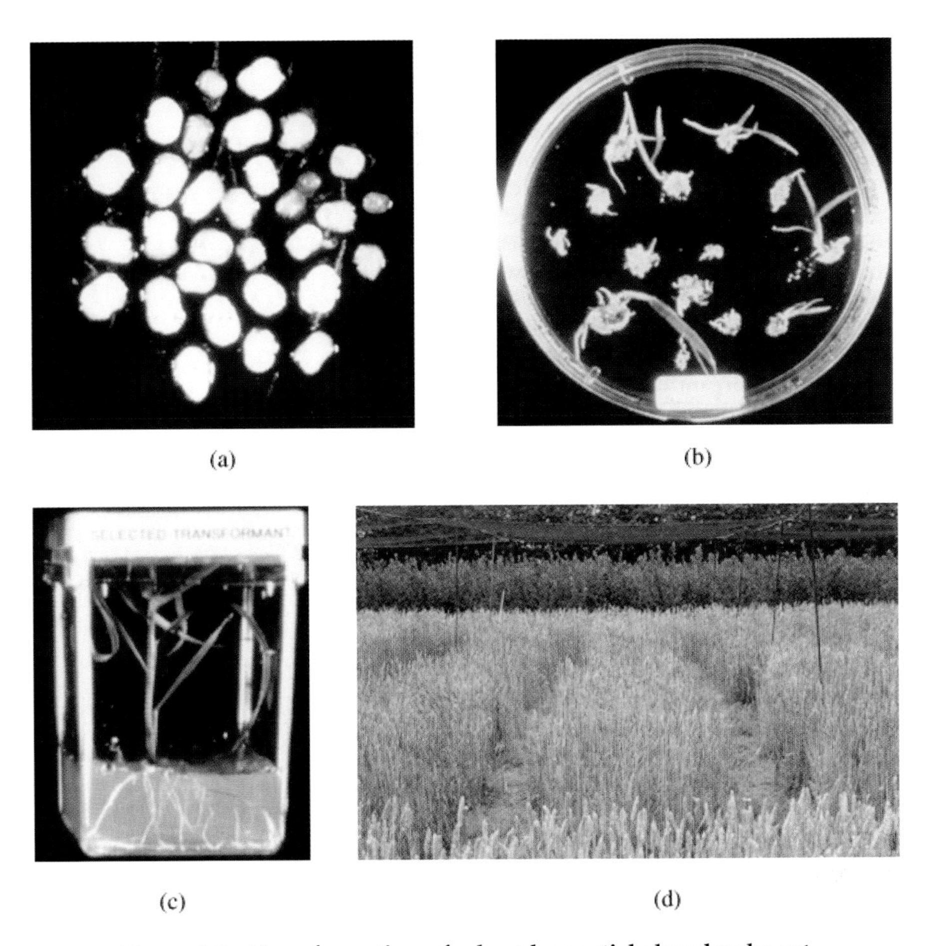

Figure 2.3 Transformation of wheat by particle bombardment.

a. Wheat embryos are the targets for bombardment with tiny gold particles coated with DNA. Some of the cells within the embryos take up the DNA and incorporate it within their own DNA.
b. The genetically modified cells are induced to form callus material and then shoots by the addition of plant hormones.
c. Shoots are placed on a different medium lacking hormones and begin to form roots.
d. Once whole plants have been produced they can be treated in the same way as non-GM wheat plants. This picture shows a field-trial of GM wheat at Long Ashton Research Station near Bristol, UK.

Pictures kindly provided by Pilar Barcelo (a–c) and Peter Shewry (d).

2.7 *Agrobacterium*-mediated Transformation Without Tissue Culture

The methods of plant transformation through the infection of explants or protoplasts with *Agrobacterium tumefaciens* and the regeneration of intact plants is relatively straightforward with many plant species, including major crops such as soybean, potato and oilseed rape. It does have the drawback, however, along with the direct gene transfer methods, that the regeneration of plants from single cells often gives rise to mutation. Plants carrying harmful mutations have to be screened out before the GM plant is incorporated into a breeding programme. If the genetic modification is being made in order to investigate the function of a particular gene for research purposes, the effects of the genetic modification may be difficult to distinguish from the effects of random mutations.

For this reason, and in order to speed up the process, methods of plant transformation that do not require tissue culture have been developed. The most successful of these is floral dip transformation. This was developed by Georges Pelletier, Nicole Bechtold and Jeff Ellis using the model plant, Arabidopsis. Plants at the early stages of flowering are placed in a suspension of *Agrobacterium tumefaciens* in a vacuum jar, a vacuum is applied to remove air surrounding the plant tissue and allow the bacteria to come into contact with it, and the plants are grown to seed. Typically, approximately 1% of the seeds are genetically modified. This method is now widely used in basic research using Arabidopsis and has been adapted with some success for use with other plant species, including soybean and rice.

2.8 Selectable Marker Genes

A limitation to all of the plant genetic modification techniques described above is that only some of the cells in the target tissue are genetically modified, irrespective of the method of DNA transfer. It is, therefore, necessary to select out all of the cells or regenerating plants that are not modified and this requires that the presence of the gene of interest can be selected for, in other words the gene must make the host plant distinguishable in some way from plants that do not carry it. Alternatively, the gene of interest must be accompanied by at least one other gene that acts as a selectable marker. Most commonly, the regeneration of GM plants is carried out in the presence of a selective agent, tolerance of which is imparted by the introduced gene or, more often, the accompanying selectable marker gene.

In practice, selectable marker genes make the transformed cells and GM plant resistant to an antibiotic (Figure 2.4a) or tolerant of a herbicide (Figure 2.4b). Those genes that are used confer resistance to kanamycin, geneticin or paromycin (collectively known as aminoglycosides) or to hygromycin, none of which have critical, if any, use in medicine. Kanamycin, geneticin and paromycin resistance is imparted by a gene called *NPTII* (neomycin phosphotransferase) while hygromycin resistance is imparted by one of *HPT*, *HPH* or *APH-IV* (hygromycin phosphotransferase). Antibiotic resistance genes are widespread in nature, and all of these genes were obtained originally from the common gut bacterium *Escherichia coli*. The *NPTII* gene is probably the most widely used selective marker gene in the genetic modification of dicotyledonous plants. However, it is not generally used in the genetic modification of cereals because kanamycin is not toxic enough to cereal cells and plants.

The use of antibiotic resistance marker genes in plant biotechnology is a controversial issue and it is now generally avoided in the production of GM crops for food and feed use. However, antibiotic resistance marker genes remain an extremely useful tool in plant genetic modification for research purposes. This issue is discussed in more detail in Chapter 5. In contrast, GM crops with herbicide tolerance are now widely used commercially (Chapter 3). The use of herbicide tolerance genes as selectable markers to accompany other genes of interest was developed because the removal of non-GM cells and regenerating plants with antibiotics was found to be unreliable with cereals. Herbicide tolerance genes work either by encoding an

(a)

(b)

Figure 2.4 Antibiotic resistance and herbicide tolerance marker genes.

a. A genetically modified tobacco plant (right) carrying a gene that imparts resistance to an antibiotic, kanamycin, and an unmodified plant (left) growing in the presence of the antibiotic.
b. A genetically modified wheat plant (right) carrying a gene that imparts tolerance of a herbicide, bialaphos, and an unmodified plant (left) in the presence of the herbicide. Picture kindly provided by Pilar Barcelo.

insensitive version of a protein that is targeted by the herbicide, or by encoding an enzyme that breaks the herbicide down.

The most widely used herbicide tolerance genes for research purposes are the *BAR* or *PAT* (phosphinothrycin acetyl transferase) genes, from the bacteria *Streptomyces hygroscopicus* and

Streptomyces viridochromogenes. They impart tolerance to herbicides based on phosphinothrycin, including gluphosinate and bialaphos, which inhibit the enzyme glutamine synthetase. Glutamine synthetase is required for the incorporation of inorganic nitrogen into amino acids, and therefore for protein synthesis. The herbicide tolerance genes encode enzymes that convert the herbicide to a non-toxic compound; phosphinothrycin acetyl transferase, for example, converts phosphinothrycin to acetylphosphinothrycin (Figure 2.5).

This selection method works for all of the cereals that have been transformed to date and has also been used in some commercial herbicide-tolerant GM crop varieties (Chapter 3). Less widely used for research but also with applications in commercial GM crop varieties are genes that impart tolerance of glyphosate, such as the *EPSPS* (5′-enolpyruvylshikimate phosphate synthase) and *GOX* (glyphosate oxidoreductase) genes from *Agrobacterium*. The enzyme

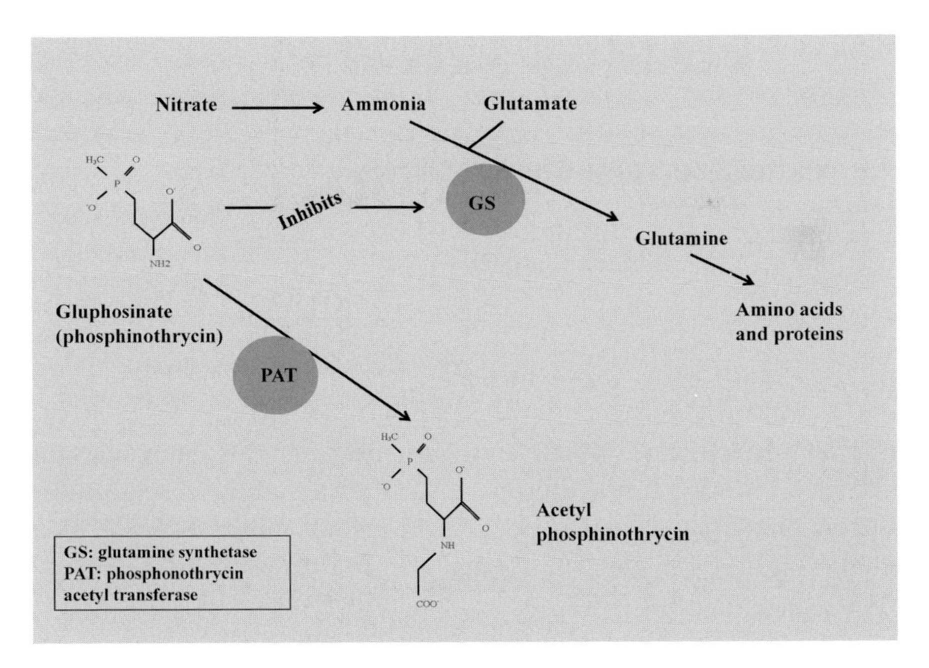

Figure 2.5 The action of gluphosinate on amino acid synthesis and the detoxifying action of phosphinothricine acetyl transferase (PAT).

EPSPS catalyses the formation of 5-enolpyruvoylshikimate 3-phosphate (EPSP) from phosphoenolpyruvate (PEP) and shikimate 3-phosphate (S3P). This reaction is the penultimate step in the shikimate pathway (Figure 2.6), which results in the formation of chorismate, which in turn is required for the synthesis of many aromatic plant metabolites including the amino acids phenylalanine, tyrosine and tryptophan. The shikimate pathway is not present in animals, which have to acquire phenylalanine, tyrosine and tryptophan in their diet (hence they are referred to as essential amino acids); for this reason, glyphosate has low toxicity for animals. The EPSPS gene from *Agrobacterium tumefaciens* encodes an EPSPS that is not affected by glyphosate, so this system bypasses the action of the herbicide, rather than breaking the herbicide down (Figure 2.6). On the other hand, glyphosate oxidoreductase, an enzyme encoded by the GOX gene, also from *Agrobacterium tumefaciens*, breaks glyphosate down to aminomethylphosphonic acid.

Finally, there are genes that impart tolerance of sulfonylurea compounds such as chlorsulfuron, including the *ALS* (acetolactate synthase) gene from maize.

2.9 Visual/Scoreable Marker Genes

A marker gene that encodes a visual or scoreable product may be used alongside a selectable marker gene in order to allow GM cells to be visualised. Examples of genes used in this way are a bacterial gene, *UidA* (commonly called Gus), a jellyfish green fluorescent protein (GFP) gene and a luciferase (*Lux*) gene. The *UidA* gene encodes an enzyme called β-glucuronidase. When supplied with 4-methyl umbelliferyl glucuronide as substrate this enzyme will produce 4-methyl umbelliferone, a substance that fluoresces under ultra-violet light at a wavelength of 365 nm, emitting light at a wavelength of 455 nm. The intensity of fluorescence can be measured using a fluorimeter, giving an indication of the amount of product and therefore the amount of enzyme that is being produced and the activity of the gene. Plant tissue is ground up in this experiment to allow the enzyme and substrate to come together.

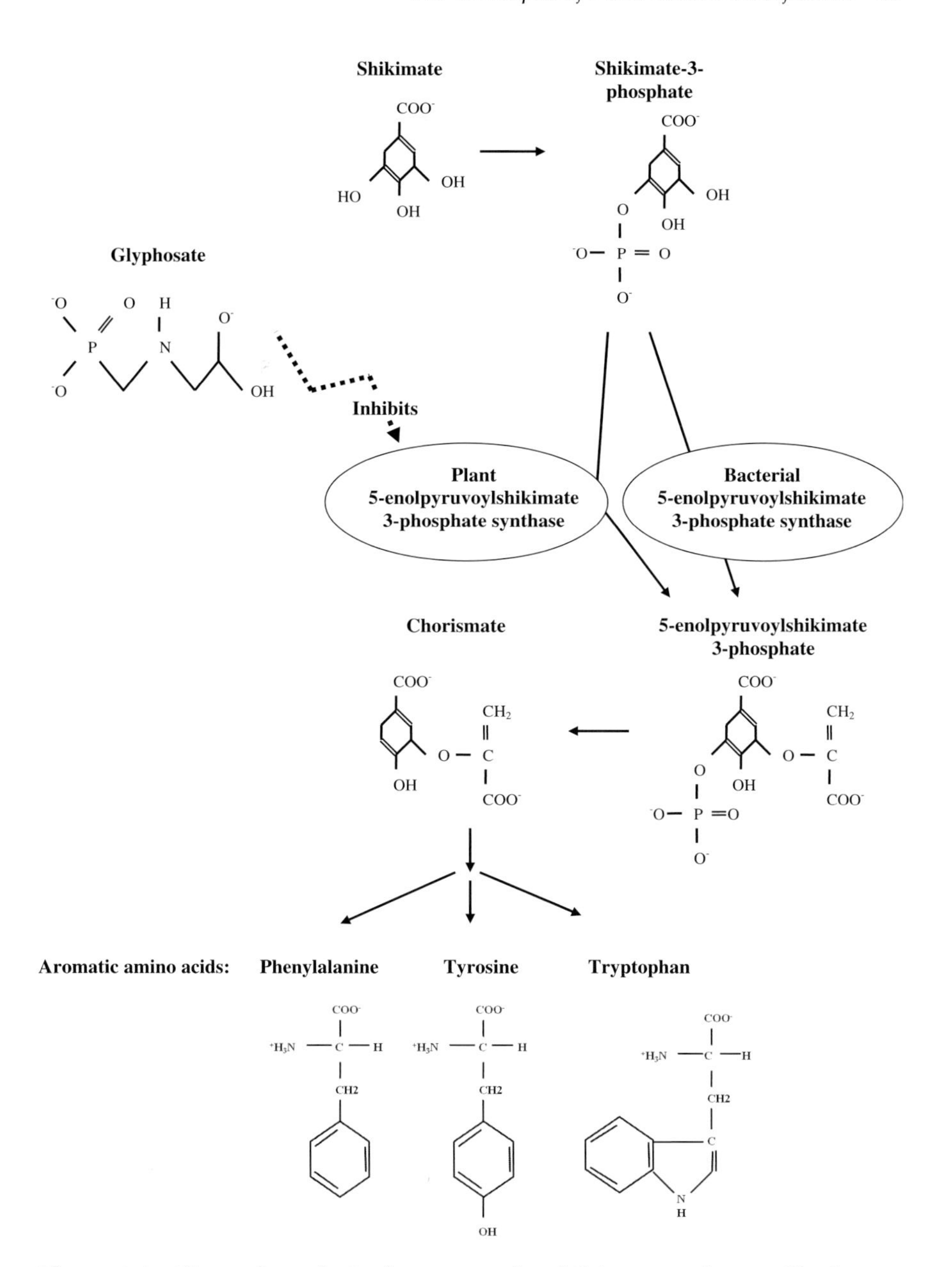

Figure 2.6 The action of glyphosate on the shikimate pathway. Glyphosate inhibits plant 5-enolpyruvoylshikimate 3-phosphate synthase (EPSPS), but does not affect the bacterial form of the enzyme.

An alternative substrate, X-Gluc (5-bromo-4-chloro-3-indolyl- β-D-glucuronide) can be used to penetrate thin sections of plant tissue. The β-glucuronidase enzyme produces a blue product from this substrate and the blue colour can be seen when the tissue is examined through a microscope, allowing the exact location of gene activity to be pinpointed (Figure 2.7a).

The *UidA* gene has proved extremely valuable as a research tool. Its main drawback is that it is destructive; the plant tissue is killed by the assay. The products of the luciferase and GFP genes, on the other hand, can be visualised without killing the plant tissue and have little or no toxicity for the plant cell. The favoured luciferase system actually requires two genes, *LuxA* and *LuxB*, from a bacterium, *Vibrio harveyi*. Activity of the two genes and the combining of the two gene products results in bioluminescence within the tissue and this can be detected and measured by light-sensitive equipment.

The GFP gene comes from an intensely luminescent jellyfish, *Aequorea victoria*. Light-emitting granules are present in clusters of cells around the margin of the jellyfish umbrella and contain two proteins, aequorin, which emits blue-green light, and green fluorescent protein (GFP), which accepts energy from aequorin and re-emits it as green light. GFP fluoresces maximally when excited at 400 nm and it is visualised in GM plants using a fluorescence stereomicroscope (Figure 2.7b).

While they are undoubtedly powerful tools in GM research, the presence of visual/scoreable marker genes in a commercial crop variety is not essential and it is accepted that they should be avoided when producing GM plants for food or animal feed.

2.10 Design and Construction of Genes for Introduction into Plants

The coding sequence of a gene that is sourced from a bacterium, an animal or a different plant species will usually be translated properly to make a protein in a GM plant. However, the regulatory sequences within a gene that determine when and where in the

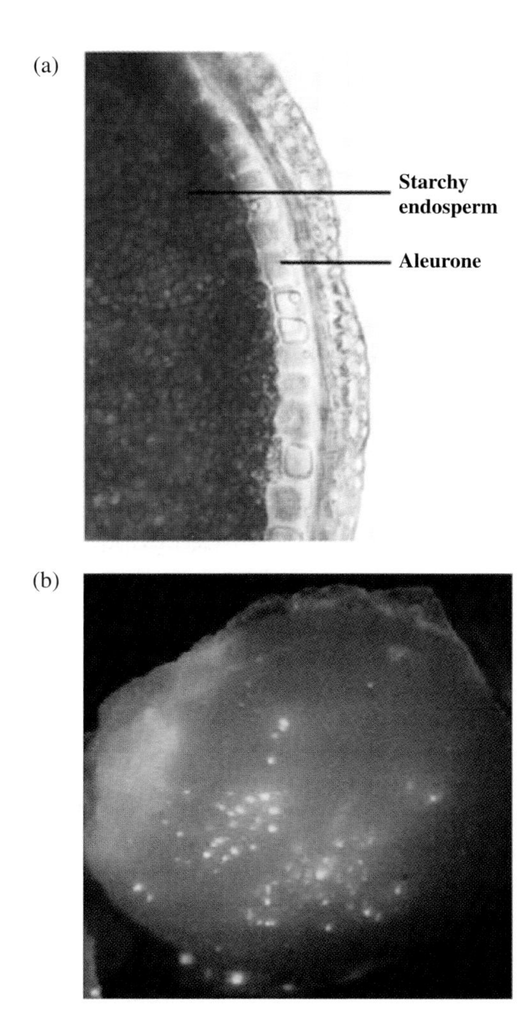

Figure 2.7 *UidA* (Gus) and green fluorescent protein (GFP) visible marker genes.

a. The *UidA* gene is bacterial in origin and encodes an enzyme called β-glucuronidase. When supplied with the appropriate substrate this enzyme will produce a blue product, allowing the exact location of gene activity to be pinpointed. In this picture the gene is under the control of a wheat gene promoter that is active only in the seed endosperm (the major seed storage tissue in cereals). The picture contrasts the dark-stained endosperm cells with the unstained cells of the surrounding tissue called the aleurone.
b. The product of the GFP gene (from the jellyfish *Aequorea victoria*) fluoresces when excited at 400 nm and is visualised using a fluorescence stereomicroscope. This picture shows bright fluorescent foci in a wheat embryo that has been bombarded with a GFP gene controlled by a wheat promoter that is active in this tissue.
Picture kindly provided by Sophie Laurie.

organism the gene is active are less likely to be recognised the more distantly related the source organism is from the GM plant. Hence, a bacterial gene introduced unchanged into a GM plant will not be active at all, and *vice versa*. This means that the coding region of the gene to be introduced into a plant is usually spliced together with regulatory sequences that will work in the plant to make what is called a chimaeric gene.

Figure 2.8 shows a schematic diagram of a chimaeric gene comprising a promoter from a wheat seed protein gene (*Glu-D1x*) attached to the coding sequence from the bacterial gene, *UidA*, and the terminator from the *Agrobacterium* nopaline synthase gene (*Nos*). As described in the previous section, the *UidA* coding sequence encodes an enzyme called β-glucuronidase that produces a blue pigment from a colourless substrate. The production of the blue product when the substrate is supplied shows that the enzyme is present and that the gene must be active. The *Nos* gene is one of those that is introduced into a plant during infection by wild-type *Agrobacterium tumefaciens* and has evolved to function in plant cells. The terminator sequence ensures that RNA from the introduced gene is processed properly. The other regulatory sequences of the chimaeric gene are present in the promoter from the wheat gene.

When wheat plants are genetically modified with this gene, the gene is active in the same tissue and at the same time as the gene from which the promoter came. That is specifically in the major storage compartment of the seed (the endosperm) in the mid-term

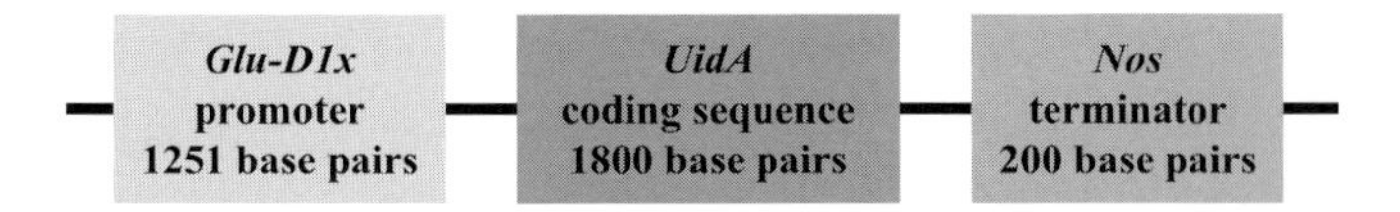

Figure 2.8 Schematic representation of a chimaeric gene comprising the coding region of the *UidA* gene from a bacterium linked to a wheat gene promoter (*Glu-D1x*) that is active only in the seed endosperm and a terminator sequence from an *Agrobacterium tumefaciens* nopaline synthase (*Nos*) gene that functions in plants. Part of a section of a seed from a plant transformed with this gene is shown in Figure 2.7a.

of seed development. This can be shown by sectioning seeds from the GM wheat plants and incubating the sections in a medium containing the substrate for β-glucuronidase, as shown in Figure 2.7a. A blue colour develops in the endosperm tissue from mid-development seeds but not in sections taken from anywhere else in the plant.

2.11 Promoter Types

The *Glu-D1x* promoter is known as a tissue-specific and developmentally regulated promoter, being active in only one tissue type in the plant at a specific developmental stage. There are many different promoters of this type now available for plant biotechnology, and they are active in many different tissues and even cell types. For example, Figure 2.9 shows β-glucuronidase activity in the roots, anthers and egg sacs of three different GM wheat plants, each containing the *UidA* gene under the control of a different promoter. It is an excellent demonstration of the control over the activity of a gene that is available to a biotechnologist. Such control is not possible with other methods in plant breeding. This tight control over promoter activity may be lost if the promoter is used in a different species, however, and this has to be tested before the promoter is used in commercial applications.

There are two other types of promoter available to the plant biotechnologist. The first type is known as constitutive, meaning that these promoters are active everywhere in the plant all of the time. In GM research, constitutive promoters are used most often to drive expression of selectable marker genes. However, the fact that they are expressed everywhere in the plant has also made them the promoter of choice for use in genes that impart herbicide tolerance or insect resistance in commercial GM crop varieties.

The most widely used constitutive promoter is derived from *Cauliflower mosaic virus*: it is called the *Cauliflower mosaic virus* 35S promoter, CaMV35S promoter for short, and was first characterised by Nam-Hai Chua and colleagues at Rockefeller University in the 1980s. *Cauliflower mosaic virus* (CaMV) infects plants of the

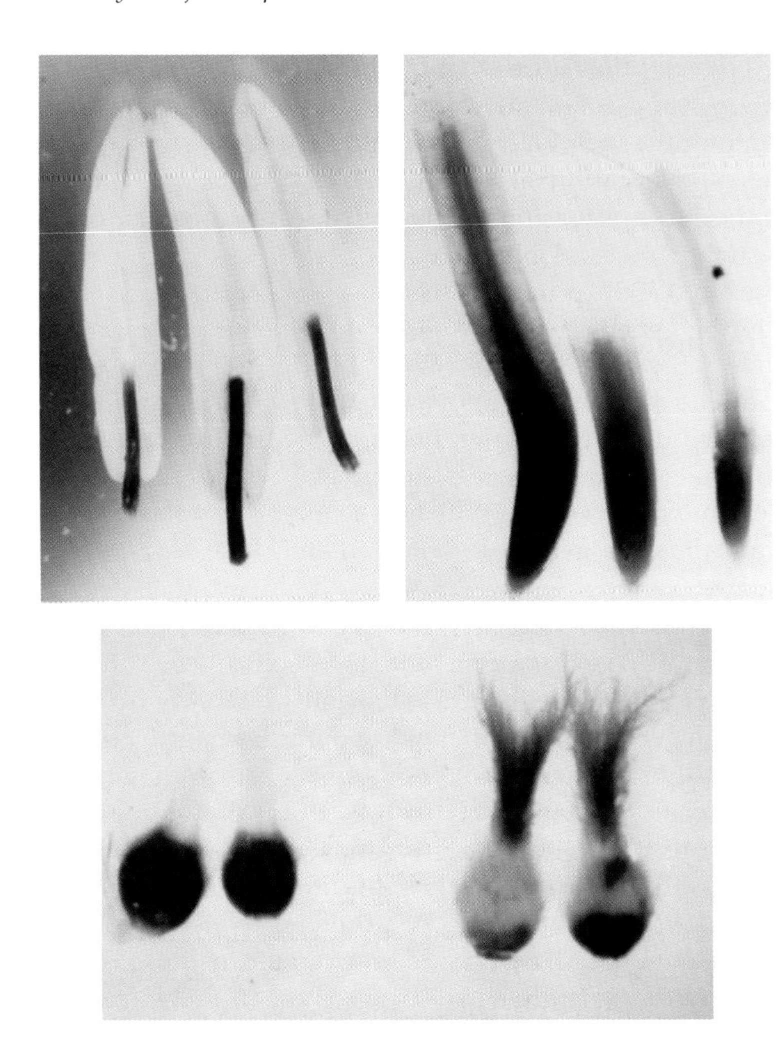

Figure 2.9 Blue staining showing *UidA* gene activity in wheat plants that have been engineered with the gene under the control of different wheat promoters. Top left: activity in anther stalks. Top right: activity in different parts of the root. Bottom: activity in different regions of the egg sacs.
Pictures kindly provided by Pilar Barcelo.

cabbage family. Its DNA is a circular molecule of about 8,000 base pairs (8 kb) and when it enters a host cell it is transcribed in its entirety to give a 35 S RNA molecule or in part to give a 19 S RNA molecule. The CaMV35S promoter is therefore responsible for transcription of the entire viral genome. Note that the S in the name refers to the sedimentation coefficient, which characterises the behaviour of a molecule or particle in sedimentation processes such as centrifugation; although not dependent solely on the mass of the molecule, it was one of the first methods developed for distinguishing between DNA and RNA molecules of different sizes.

The CaMV35S promoter has been used particularly successfully in the genetic modification of dicotyledonous plants. It also works in monocotyledonous plants, including cereal crop species, but has largely been replaced for cereal work with promoters that are derived from cereal genes. These include a promoter for a maize gene, *Ubi*, encoding a protein called ubiquitin, and a rice gene, *Act1*, encoding a protein called actin. There has been some controversy over the use of the CaMV35S promoter (although not in the plant science community) because of its viral origins, even though the *Cauliflower mosaic virus* only infects plants and the CaMV35S promoter represents only a small part of the viral genome and could not in itself be infective.

The third type of promoter that a biotechnologist can use is described as inducible. Promoters of this type are not active until they are induced by something such as attack by a pathogen, grazing or application of a chemical. There is some interest in using pathogen- or grazing-induced promoters in GM crops that are being modified to resist pathogen or insect attack. The resistance gene introduced into the plant would then only be active when it was needed. Chemically induced promoters are used in GM research so that a gene can be switched on and off in order to determine its function. Such promoters are not yet being used in commercial GM crops.

2.12 The Use of GM to Characterise Gene Promoters

A common misconception regarding plant genetic modification is that it is entirely concerned with producing new crop varieties for agriculture and is the preserve of big business. In fact, the techniques were developed in the public sector, largely in the USA and Europe, and have been extremely valuable in basic plant genetic research. The types of experiments to which it has been applied include analyses of gene promoter activity, functional characterisation of regulatory elements within gene promoters, the determination of gene function, studies on metabolic pathways and analyses of protein structure and function.

The analysis of gene promoter activity is one way of finding out when and where in a plant a gene is active. The promoter of the gene is spliced to a reporter gene such as *UidA*, GFP or luciferase and the resulting chimaeric gene, known as a reporter gene construct, is introduced into a plant by genetic modification. The level and location of promoter activity can be visualised by testing for the presence of the protein produced by the reporter gene (see Figures 2.7 and 2.9).

Further experiments can be undertaken to determine what parts of the promoter are important in its activity. This process often begins with an analysis of different promoters that have similar activity in order to find short regions of DNA (elements) that have the same sequence of base pairs. These elements are likely to play a role in controlling the activity of the promoter and are called regulatory elements. The function of potential regulatory elements can then be tested experimentally.

A good example of this was an analysis of the promoters of seed storage protein genes of wheat, barley, rye and maize. As their name suggests, these genes encode proteins that function as part of the storage reserve in the seed, providing nutrition for the developing seedling after germination. They have evolved into a large gene family, but are believed to have arisen from a single ancestral gene, and their activity is controlled in a co-ordinate manner. They are subject to tissue-specific and developmental regulation, being expressed exclusively in the major storage compartment of the seed, the starchy endosperm, during mid and late seed development. They are also subject to nutritional regulation, responding sensitively to the availability of nitrogen and sulphur in the grain.

Since they have a common ancestry and show similar patterns of expression, it was to be expected that these genes would have regulatory sequences in common, and this turned out to be true for many of them. One group was found to contain a conserved element, 29 base pairs long, positioned around 300 base pairs upstream of the coding sequence of the gene. This was one of the first plant regulatory elements to be characterised. It was identified by Brian Forde at Rothamsted in the UK and was first called the –300 element, subsequently the prolamin box. It has the nucleotide sequence: TGACATGTAAAGTGAATAAGATGAGTCAT.

A regulatory role for the prolamin box was established experimentally by particle bombardment of cultured barley seed endosperms with promoter/*UidA* reporter gene constructs containing different regions of the promoter. These experiments also showed that the prolamin box could be subdivided into two separate elements, one (the E box) conferring tissue-specificity, the other (the N box) reducing activity of the gene at low nitrogen levels and increasing it when nitrogen levels were adequate. All of this control is imparted by the 29 base pair element.

Regulatory elements that have a positive effect on gene expression are called enhancers, while those that have a negative effect on gene expression are called silencers. The E box is an example of a tissue-specific enhancer because it causes a large increase in gene expression specifically in the seed endosperm, while the N box can act as an enhancer or a silencer, depending on nitrogen availability.

2.13 Gene Over-Expression and Silencing

The use of genetic modification in the analysis of gene function is based on finding out what happens when a gene is switched off when it should be active (gene silencing) or what happens if the gene is over-expressed to make more of a protein or to make it in a tissue type or at a developmental stage where it would not normally be present. Over-expression can be achieved by introducing additional copies of a gene unchanged, or by introducing a modified version of the gene in which the coding region is spliced to a more powerful promoter.

Several methods have been developed for silencing genes in plants using genetic modification, but they are now known to involve essentially the same mechanism. Historically, the first method to be developed was the antisense technique, in which a chimaeric gene is produced using part of the gene of interest spliced in reverse orientation downstream of a promoter. The promoter may derive from the same gene, but usually it is a more powerful one. When this chimaeric gene is re-inserted into the plant, it produces RNA of the reverse and complementary sequence of that produced by the endogenous gene. This so-called antisense RNA interferes with the accumulation of sense RNA from the target gene, preventing the sense RNA from acting as a template for protein synthesis. This method was developed by Don Grierson and colleagues at the University of Nottingham in collaboration with a team led by Wolfgang Schuch at what was then Imperial Chemical Industries (ICI) (the seed division at ICI was one of the sections of the company that was subsequently split off as Zeneca and after a number of re-organisations and mergers became Syngenta).

Grierson and Schuch also developed the co-suppression technique, in which one or more additional copies of all or part of a gene in the correct orientation is introduced into a plant by genetic modification. This, of course, is the technique used for over-expression of a gene but in some cases it leads to gene silencing. Gene silencing by this method requires the introduced gene to be extremely similar or identical to the native gene, so can be avoided if over-expression is the desired outcome by using a gene from a not too closely related species.

Both co-suppression and antisense gene silencing have been used to produce genetically modified plants in which the trait is stably inherited, and both techniques have found commercial application in the extension of fruit shelf-life, most famously that of the tomato (see Chapter 3). There may be more than one silencing mechanism involved in both of these techniques, but the major one appears to operate post-transcriptionally, causing degradation of the RNA molecule before it can act as a template for protein synthesis. Post-transcriptional gene silencing (PTGS) turns out to be a defence mechanism against virus infection; indeed, some plant viruses have genes that suppress it. It involves the production of small, antisense RNAs. Investigations into PTGS by David Baulcombe and Andrew Hamilton, then at the Sainsbury Laboratory in Norwich, led to the development of the third method for gene silencing by genetic modification, RNA interference (RNAi), and the discovery that PTGS was the unifying mechanism for the antisense, co-suppression and RNAi techniques.

In RNAi, a plant is genetically modified to synthesise a double-stranded RNA molecule derived from the target gene. This has been done in other systems by splicing part of a gene between two opposing promoters and re-introducing it into an organism. It has been achieved in plants using gene constructs in which part of the gene is spliced sequentially in a head-to-tail formation downstream of a single promoter. This causes the production of an RNA molecule that forms a hairpin loop (hpRNA); this molecule is cleaved into short, double-stranded RNA molecules by an enzyme called Dicer that is naturally present in the cell. These short RNA molecules are called short interfering RNAs (siRNAs). The siRNAs are unwound into two single-stranded molecules, one of which (the passenger strand) is degraded, while the other (the guide strand) is incorporated into a RNA-induced silencing complex (RISC). The guide strand pairs with the complementary sequence of messenger RNA from the target gene and induces cleavage by another enzyme, Argonaute, which is present in the RISC.

The fact that PTGS is a natural response to viral infection has led to the development of virus-induced gene silencing (VIGS) to suppress gene expression transiently. This involves the production of modified plant viruses carrying nucleotide sequences corresponding to the host gene to be silenced. Infection leads to synthesis of viral double-stranded RNA, as an intermediate step in the normal viral replication process. This activates the plant's antiviral RNA silencing pathway, resulting in degradation of the messenger RNA from the target gene. Although still a relatively new technique, VIGS is proving to be a powerful tool in plant genetic research.

3 THE USE OF GM CROPS IN AGRICULTURE

3.1 Why Use Genetic Modification (GM) in Plant Breeding?

The graph shown in Figure 1.7 gives an indication of how successful plant breeding has been over the last century. So why do plant breeders need GM? The answer is that GM allows plant breeders to do some things that are not possible by other techniques. That does not mean to say that GM will replace older techniques in plant breeding, far from it, but GM is undoubtedly a powerful new tool for plant breeders to use. The advantages that GM has over other techniques are as follows:

1. It allows genes to be introduced into a crop plant from any source. Biotechnologists can select a gene from anywhere in nature and with the modifications described in Chapter 2 make a version of it that will be active in a crop plant.
2. It is relatively precise in that single genes can be transferred. In contrast, conventional plant breeding involves the mixing of tens of thousands of genes, many of unknown function, from different parent lines, while radiation and chemical mutagenesis introduce random genetic changes with unpredictable consequences.
3. Genes can be designed to be active at different stages of a plant's development or in specific organs, tissues or cell types (see Chapter 2).

4. Specific changes can be made to a gene to change the properties of the protein that it encodes.
5. The nature and safety of the protein produced by a gene can be studied before the gene is used in a GM programme (see Chapter 4).

GM also has some disadvantages. A successful GM programme requires background knowledge of a gene, the protein that it encodes and the other genes and proteins that interact with it. This requires a significant investment of time and money compared, for example, with the generation of random mutations. Most significantly, though, GM varieties have to undergo much more detailed analysis and testing than new non-GM varieties, particularly in Europe. This probably explains why the few GM crop varieties available in Europe are essentially spin-offs from varieties developed elsewhere. The barriers to developing GM crop varieties specifically for the European market are too great. Even outside Europe, the GM crops that are currently available are based on a handful of strategies, although the number of different GM applications is growing.

Typically, development of a new GM variety takes at least ten years and is estimated to cost $100 million (Figure 3.1). The cost escalates through the process, which means that it will always be tempting for chief executives to call a halt, and regulatory compliance adds significantly to the cost. Establishing a market is also a problem and some GM varieties have come and gone without doing so.

On the other hand, some of the GM crops that have made it onto the market have been staggeringly successful. Data on the global use of GM crops has been compiled for several years by Clive James of the International Service for the Acquisition of Agri-Biotech Applications (ISAAA; www.isaaa.org). It is almost impossible to check the ISAAA's data for some countries, but for countries where independent data is available from other sources, the USA for example, the datasets are consistent. According to the ISAAA, the worldwide area of land planted with GM crops in 2009

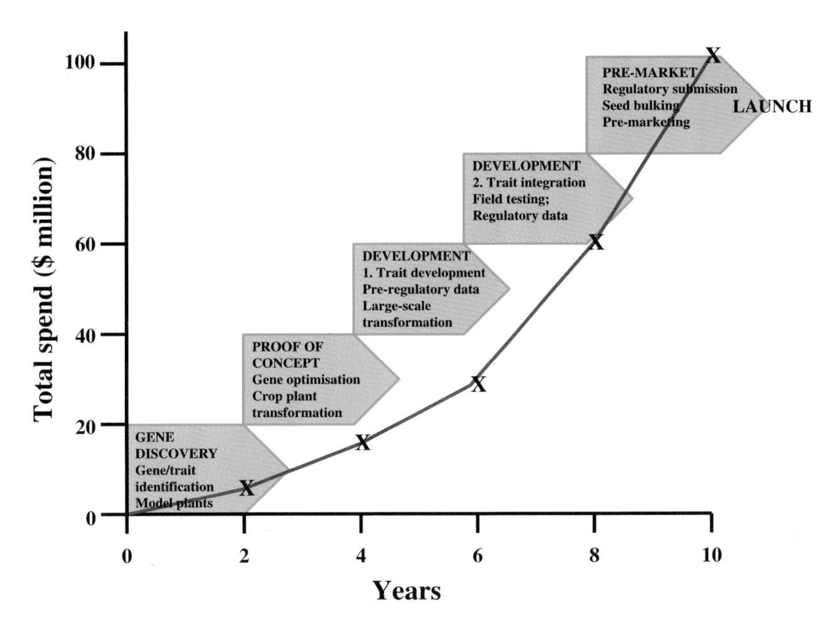

Figure 3.1 Flow-chart showing the stages of development of a new GM crop variety, and graphical representation of the typical time and expense involved.

was 134 million hectares (330 million acres), approximately 9% of total world agriculture. More than three-quarters (77%) of global soybean production was GM, with figures of 49%, 26% and 21% for cotton, maize and oilseed rape, respectively. Minor crops with some GM varieties being grown commercially included papaya, squash, tomato and tobacco. GM crops were grown in 25 countries (Table 3.1), with the USA, Argentina, Brazil, India, Canada, China, Paraguay and South Africa all planting more than a million hectares of GM crops.

There are some notable absentees from these lists. There is currently no GM wheat being grown commercially on a significant scale, for example, and very little GM rice. China has just approved some GM rice varieties for cultivation but it is likely to be 2012 before they are made available to farmers. Wheat may take longer: it looked like GM wheat was going to be launched by Monsanto in 2004 but the company performed an abrupt about-face and pulled

Table 3.1 GM crop areas in the 25 countries growing GM crops commercially in 2009.

Country	GM crop area (thousand hectares)
USA	64000
Brazil	21400
Argentina	21300
India	8400
Canada	8200
China	3700
Paraguay	2200
South Africa	2100
Bolivia	800
Uruguay	800
Philippines	500
Australia	200
Burkina Faso	100
Mexico	100
Spain	100
Chile	>50
Colombia	>50
Costa Rica	>50
Czech Republic	>50
Egypt	>50
Honduras	>50
Poland	>50
Portugal	>50
Romania	>50
Slovakia	>50
Total	**134000**

out of wheat research and development altogether. No other company looks set to launch a GM wheat variety and although Monsanto has now renewed its interest in wheat it does not appear to have any GM wheat varieties ready for commercialisation. As far as the list of countries goes, there are very few from Europe or Africa. This is discussed in more detail later.

The first GM trait to come on to the market was delayed ripening, a trait that was initially and most successfully developed in tomato. The first traits to come on to the market in the major commodity crops were herbicide tolerance and insect resistance, and these remain by far the most successful traits that have been introduced into crop plants by GM. Other traits that have been used successfully include modification of oil content and virus resistance.

3.2 Slow-Ripening Fruit

Fruit ripening is a complex molecular and physiological process that brings about the softening of cell walls, sweetening and the production of compounds that impart colour, flavour and aroma. The process is induced by the production of a plant hormone called ethylene. The problem for growers and retailers is that ripening is followed sometimes quite rapidly by deterioration and decay and the product becomes worthless. Tomatoes and other fruits are, therefore, usually picked and transported when they are unripe. In some countries they are then sprayed with ethylene before sale to the consumer to induce ripening. However, fruit that is picked before it is ripe has less flavour than fruit picked ripe from the vine.

Biotechnologists therefore saw an opportunity in delaying the ripening and softening process in fruit. If ripening could be slowed down by interfering with ethylene production or with the processes that respond to ethylene, fruit could be left on the plant until it was ripe and full of flavour but would still be in good condition when it arrived at the supermarket shelf. Various strategies based on this principle are now being pursued with many different fruits and the technology has the potential not only to improve the produce of western farmers but also to enable farmers in tropical countries to sell fruit to customers in Europe and North America. However, although slow-ripening GM varieties of tomato and papaya are grown in China, the GM tomato varieties that have been marketed in the west have been withdrawn for one reason or another, and the application of the technology in other fruits is still some years away from commercialisation.

The first commercial varieties of GM tomatoes were engineered to inhibit the process by which fruits soften as they ripen. The GM fruits change colour and acquire aroma but are less prone to bruising and deteriorate and decay much more slowly than normal. Fruit softening is brought about by enzymes that degrade the complex carbohydrates in the cell walls within the fruit. These carbohydrates include cellulose, which comprises linear chains of glucose molecules (a type of glucan), and pectin, which is comprised of a complex group of polysaccharides based on galacturonic acid, an oxidised form of galactose, and rhamnose, which is a deoxyhexose, with various sugar side-chains. The enzymes involved include cellulases, which break down cellulose, and polygalacturonase (PG) and pectin methylesterase (PME), which are involved in the breakdown of pectin.

Two competing groups developed tomatoes in which PG activity had been reduced in the fruit by gene suppression (see Chapter 2) at approximately the same time: Calgene in the USA and Zeneca (as it was then known) in collaboration with Don Grierson's group at the University of Nottingham in the UK. The Calgene product was the Flavr Savr tomato, the first GM product on the market. It was introduced in 1994 but was not a commercial success and was withdrawn after less than one year. Calgene was subsequently acquired by Monsanto, who have so far not pursued the technology further.

Zeneca chose to introduce the trait into a tomato used for processing and this proved to be much more successful. These tomatoes had a higher solid content than conventional varieties, reducing waste and processing costs in paste production and giving a paste of thicker consistency. This product went on the market in many countries and proved very popular in the UK where over two million cans of it were sold from its introduction in 1996 until 1999 when most retailers withdrew it in response to anti-GM hostility.

Another method for slowing down fruit ripening is to engineer a plant so that the fruit does not produce ethylene. The advantage of this strategy is that the fruit develops to the point where it would normally start to ripen and then stops, allowing the farmer to wait until all of the fruit has ripened and then harvest it all at once. Ripening is then induced by spraying the fruit with ethylene.

Tomatoes of this type have been developed by suppressing the gene that encodes the enzyme aminocyclopropane-1-carboxylic acid (ACC) synthase. ACC synthase is responsible for one of the steps in ethylene synthesis and reducing levels of ACC decreases ethylene production dramatically. A tomato variety of this type was developed by a company called DNA Plant Technologies and marketed in the USA in the 1990s under the trade name 'Endless Summer'. However, the variety was withdrawn from sale because of disputes over patenting.

An alternative method with a similar outcome is to add a gene that encodes an enzyme called ACC deaminase. This enzyme interferes with ethylene production by breaking down ACC. Tomatoes of this type have been developed by Monsanto using a gene derived from a soil bacterium called *Pseudomonas chlororaphis* but so far have not been marketed. A third method targets another of the precursors of ethylene, S-adenosyl methionine (SAM), by introducing a gene that encodes an enzyme called SAM hydrolase, which breaks down SAM; this strategy was developed by Agritope, Inc. using a viral SAM hydrolase gene, but again has not been marketed.

3.3 Herbicide Tolerance

Herbicide tolerance was described in Chapter 2 in the context of selectable marker genes but several traits of this type have been introduced into commercial crops. Varieties carrying them have been extremely popular wherever farmers have been allowed to use them. Weed control is an essential part of all types of agriculture and in the developed world the method of choice for most farmers to achieve it is to spray fields with chemical herbicides. This has been true since the 1950s, long before the advent of genetic modification. The alternatives are labour intensive and farmers could not abandon the use of herbicides with today's labour costs and keep food prices anything close to the level that they are now. Organic farmers do not use herbicides but organic farmers in the developed world are selling into niche markets, not providing the general population with affordable food. In the United Kingdom, for example, organic farmers simply avoid growing crops like sugar beet that are particularly sensitive to weed competition.

Most herbicides are selective in the types of plant that they kill and a farmer has to use a herbicide that is tolerated by the crop that he is growing but kills the problem weeds. Commercial (non-GM) sugar beet varieties grown in the UK, for example, are tolerant of around 18 commercially available herbicides. By using a combination of different herbicides (typically eight) at different times in the season when different weeds become a problem, the farmer can protect his crop. Without weed control, sugar beet yield falls by about three-quarters and the crop is not worth harvesting.

This provides the farmer with a number of problems: the herbicide regime may be complicated; some of the herbicides may have to go into the ground before planting and weed problems that arise after the crop seed has germinated cannot be responded to; some of the herbicides involved may be toxic to humans, dangerous to handle and require protective clothing and equipment to be used; equipment and labour is required and this adds cost; finally, some herbicides persist in the soil from one season to the next, making crop rotation difficult. Many of these problems have been overcome by the introduction of GM crops that tolerate broad-range herbicides. The first of these to be introduced were Roundup-Ready soybeans, which were produced by Monsanto and have been marketed since 1996.

Roundup is Monsanto's trade name for glyphosate (Figure 2.6), a broad-range herbicide that was introduced as a commercial product by Monsanto in 1974 (pre-dating GM crops by two decades). It is now marketed under many different trade names in various agricultural and garden products. Glyphosate is non-selective and prior to its use in combination with GM plants it was used primarily to clear fields completely or to remove weeds from pathways. It does not persist long in the soil because it is broken down by micro-organisms; how long it persists depends on soil type and typically ranges from a few days to several months. This means that many farmers can clear their fields and plant a crop a few weeks later. It would be extremely rare for glyphosate to persist at effective levels from one season to the next.

Glyphosate is taken up through the foliage of a plant, so it is effective after weeds have become established. Its target is an enzyme called 5-enolpyruvoylshikimate 3-phosphate synthase (EPSPS) (Figure 2.6). This enzyme catalyses the formation of 5-enolpyruvoylshikimate 3-phosphate (EPSP) from phosphoenolpyruvate (PEP) and shikimate 3-phosphate (S3P). This reaction is the penultimate step in the shikimate pathway, which results in the formation of chorismate, which in turn is required for the synthesis of many aromatic plant metabolites. These aromatic compounds include the amino acids phenylalanine, tyrosine and tryptophan, so amongst many other things plants treated with glyphosate are unable to make proteins. The herbicide is transported around the plant through the phloem and the whole of the plant dies.

The shikimate pathway is not present in animals, so animals have to acquire its products through their diet; hence, phenylalanine, tyrosine and tryptophan are called essential amino acids, not because they are more important than other amino acids in protein synthesis but because they are an essential part of our diet. This means that glyphosate has very little toxicity to insects, birds, fish or mammals, including man. For this reason, farmers have always been very comfortable in using it.

Biotechnologists saw an opportunity to enable farmers to use glyphosate to control weeds simply and safely on growing crops by genetically modifying crop plants to tolerate the herbicide. They concentrated their efforts on EPSPS, initially trying to overcome the effect of glyphosate by engineering plants to over-produce the enzyme. Real success came with the discovery of a variant form of EPSPS (class II) in strains of the soil bacteria, *Agrobacterium tumefaciens* and *Achromobacter*. These enzymes worked efficiently but were not affected by glyphosate. The *Agrobacterium tumefaciens* gene that encodes EPSPS was isolated and inserted into soybean by particle bombardment (Chapter 2). The CaMV35S promoter (Chapter 2) was used to control the activity of the gene; the use of this constitutive promoter meant that the whole of the plant would be protected. The result was a variety of soybean that would tolerate glyphosate because while the herbicide inhibits the plant's own EPSPS enzyme it does not affect the enzyme encoded by the introduced gene.

The first glyphosate-tolerant soybean line was called 40-3-2. Line 40-3-2 was field-tested in 1992 and 1993 and gave a similar yield after glyphosate treatment to untreated non-GM varieties. The trait was given the trade name 'Roundup-Ready' and has since been incorporated into many soybean breeding programmes. Over 150 US seed companies now offer varieties carrying the trait.

Glyphosate can be applied to GM glyphosate-tolerant soybeans at any time during the season. This gives farmers flexibility in weed control and the ability to respond to a weed problem if it occurs unexpectedly. The regime that is used depends on the weed pressure in a particular area. In some cases a single application of glyphosate will control weeds throughout an entire growing season. However, the timing of the application is crucial if this is to work. Many farmers prefer to use a different herbicide before sowing in combination with a later application of glyphosate. Others use two sequential glyphosate applications.

The yield obtained with glyphosate-tolerant soybeans when they were undergoing trials was the same as for the non-GM controls. There were reports of disappointing yields in some areas when the first varieties were grown commercially, perhaps because the varieties that were available were not ideal for some locations. These reports have dried up as the trait has been bred into more varieties. There are also reports of yield increases in some areas.

The take-up of glyphosate-tolerant soybean varieties was incredibly rapid: after their introduction in 1996, their use rose to well over half of all the soybeans planted in the USA by 2001, the total area of which was approximately 30 million hectares. Adoption of GM varieties in Argentina was even more enthusiastic and reached 99% by the turn of the century. By 2009, 77% of the global soybean crop, which covered 90 million hectares, was GM, making it difficult to source non-GM soybean. Indeed, the only region where non-GM soybean can still be sourced in significant quantities is northern Brazil. Brazil authorised the cultivation of GM glyphosate-tolerant soybean in 2003 but varieties carrying the trait and suitable for the tropical climate of the north were not available at that time. That situation will not last, and it is recognised in the food and feed industries in those countries where the GM food issue remains a controversial one that non-GM soybean will not be available forever, unless it is grown to contract and a premium is paid for it. Soybean meal is a major ingredient in many processed foods and an irreplaceable source of protein for animal feed. Indeed, even in the UK, where opposition to GM foods remains strong, much of the animal feed industry now uses GM soybean.

Farmers cite simpler and safer weed control and reduced costs as the main reasons for using glyphosate-tolerant crop varieties. The exact economic impact is difficult to gauge. As glyphosate-tolerant varieties became more popular, sales of glyphosate rose while those of other herbicides, such as imazethapyr, acifluorfen, bentazon and sethoxydim, fell. The price of these herbicides was then reduced to maintain competitiveness. Overall, between 1995 and 1998 there was estimated to be a reduction of $380 million in annual herbicide expenditure by US soybean growers (source: National Council for Food and Agricultural Policy). However, farmers who used glyphosate-tolerant varieties had to pay a technology fee of $6 per acre, reducing the overall cost saving to $220 million. The technology fee did cause some resentment amongst US farmers, particularly when it was not imposed on their competitors in Argentina.

Another advantage for farmers using glyphosate-tolerant varieties is that crop rotation is made much easier. Some of the herbicides used on conventional soybean crops, such as chlorimuron, metribuzin, imazaquin and imazethapyr, remain active through to the next season and beyond. Some crops cannot be planted even three years after treatment. Maize, which is typically used in rotation with soybean, requires an eight- to ten-month gap before it can be planted. There are no such problems after glyphosate treatment because glyphosate is degraded so rapidly in the soil.

The use of glyphosate-tolerant soybean varieties has also allowed farmers to switch to a conservation tillage system, leaving the soil and weed cover undisturbed over winter. This reduces soil erosion and leaching of nitrate from the soil, where it is a fertiliser, into waterways where it is a pollutant. Conservation tillage does not require the use of herbicide-tolerant GM varieties, but the introduction of these varieties appears to have enabled or persuaded many more farmers to switch to this system.

Last but not least, the system provides farmers with peace of mind. There are even some farmers who do not use glyphosate on their Roundup-Ready soybeans at all in some seasons, preferring to use the herbicide regime that they followed before the GM variety became available. The benefit in this case is having glyphosate available 'in reserve' if a weed problem arises late in the season. This is particularly important because if weeds are not controlled at this time the soybean seed that is harvested will contain a proportion of weed seeds and buyers will pay less for it.

Glyphosate tolerance has now been engineered into many crop species and commercial varieties of cotton, oilseed rape, maize, sugar beet and fodder beet are already on the market. It is undoubtedly the most successful GM trait to be used so far. Alternatives to the glyphosate-insensitive EPSPS enzyme are being explored. Several bacteria have been found to make an enzyme called glyphosate oxidoreductase that breaks glyphosate down and detoxifies it and genes encoding this enzyme have been engineered into several crop species.

Glyphosate-tolerant varieties are not the only GM herbicide-tolerant varieties available. They face competition from varieties that are tolerant of another broad-range herbicide, gluphosinate (or glufosinate). As with glyphosate tolerance, this trait was discussed in Chapter 2 in the context of selectable marker genes and it is probably the most popular herbicide tolerance selectable marker trait used in plant genetic research. Crop varieties carrying the trait were developed by Plant Genome Systems, which was subsequently acquired by Aventis, which was then acquired by Bayer. Gluphosinate is marketed under the trade name Liberty. The gene used to render plants resistant to it comes from the bacterium *Streptomyces hygroscopicus* and encodes an enzyme called phosphinothrycine acetyl transferase

(PAT) (Figure 2.5), which detoxifies gluphosinate. Crop varieties carrying this trait have been given the trade name LibertyLink and include varieties of oilseed rape, maize, soybeans, sugar beet, fodder beet, cotton and rice. The oilseed rape variety has been particularly successful in Canada.

The other broad-range herbicide tolerance GM trait that has been used in commercial crops confers tolerance to oxynil herbicides such as bromoxynil (3,5-dibromo-4-hydroxybenzonitrile) (Figure 3.2). Bromoxynil and other oxynil herbicides inhibit photosynthesis in dicotyledonous plants by blocking electron flow during the light reaction. This causes the production of superoxide, a highly reactive free radical, also known as a reactive oxygen species (ROS), leading to the destruction of cell membranes, inhibition of chlorophyll formation and death of the plant. Tolerance is imparted by the *bxn* gene from bacterium *Klebsiella pneumoniae.* This gene encodes a nitrilase enzyme that detoxifies the herbicide (Figure 3.2).

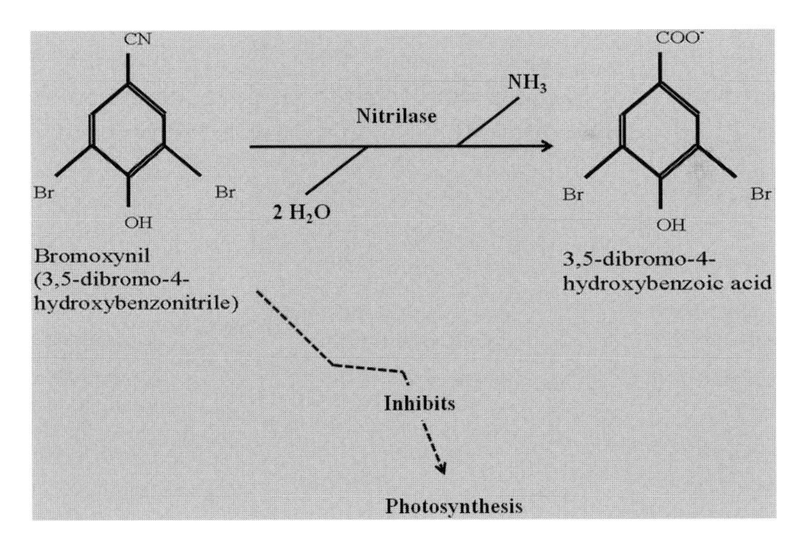

Figure 3.2 Bromoynil tolerance is engineered into plants with the introduction of the *bxn* gene from bacterium *Klebsiella pneumoniae.* This gene encodes a nitrilase enzyme that detoxifies the herbicide. For gluphosinate and glyphosate tolerance see Figures 2.5 and 2.6.

A bromoxynil-tolerant oilseed rape variety was developed by Aventis and marketed in Canada in the 1990s under the name Westar Oxy-235. This variety was withdrawn in 2002 by Bayer after the acquisition of Aventis and currently no bromoxynil-tolerant varieties of any crop are on the market. The presence of adventitious bromoxynil-tolerant oilseed rape plants in Canada has been monitored since the withdrawal of the commercial variety and is a model for the fate of GM traits in agriculture and the wider environment after they have been withdrawn. The frequency of detection has declined steadily.

One concern over herbicide-tolerant GM crops is the possibility that they could become victims of their own success through the development of tolerance to the herbicide in weed populations. This could arise through cross-pollination between the crop and related weed species, the risk of which varies with different crop species depending on whether or not they naturally out-cross and whether or not they have wild relatives in the area. This is discussed in more detail in Chapter 4. It could also arise as a result of the selective pressure on weed populations to develop resistance resulting from the widespread use of a single herbicide. Resistance to a particular herbicide becoming prevalent in weed populations would not lead to the 'superweeds' predicted by anti-GM pressure groups because farmers would control resistant weeds with different herbicides. However, it would lead to the herbicide-tolerant crop and its associated herbicide no longer being of benefit. There have been a scattering of reports of resistance to glyphosate in some areas of the USA, but this does not appear to be widespread as yet or to have dented the enthusiasm of US farmers for herbicide-tolerant GM varieties. However, it is important that other herbicide-tolerant traits are developed and used so that resistance to any one herbicide is less likely to develop.

3.4 Insect Resistance

Another universal problem for farmers all over the world is loss of their crops to insect grazing. An extreme example of this is the

Colorado beetle, which in bad years destroys 70% of the potato harvest in Eastern Europe and Russia. Farmers in developed countries usually use chemical controls and to a lesser extent biological controls (such as natural predatory species) to prevent these sorts of losses. However, these measures are expensive and the pesticides used are often toxic and difficult to handle safely, while biological controls are never 100% effective.

Organic and salad farmers have been using a pesticide called Bt for several decades. It actually consists of a soil bacterium, *Bacillus thuringiensis*, which produces a protein that is toxic to some insects. Bt pesticides are applied as powders, granules, or aqueous and oil-based liquids. Organic farmers use Bt pesticides because they are biological products that degrade rapidly. Salad farmers use them because they can be applied immediately before harvest. They have no toxicity to mammals, birds or fish and have an extremely good safety record. Disadvantages are that they are specific for insect types and do not remain effective for long after application.

The protein that *Bacillus thuringiensis* produces that is toxic to insects is called the Cry (crystal) protein. Different strains of the bacterium produce different versions of the protein, and these versions are classified into groups, CryI–CryIV. Each group is subdivided further into subgroups A, B, C etc. The different proteins are effective against different types of insects. CryI proteins, for example, are effective against the larvae of butterflies and moths, while CryIII proteins are effective against beetles.

Biotechnologists took the view that engineering crops to produce the Cry protein would be more efficient than applying the protein externally. This would overcome the problem of rapid loss of activity after application and would also have the advantage that only insects eating the crop would be affected. The Cry1A gene from several strains of *Bacillus thuringiensis* has now been introduced into crops, including cotton, sugar beet and maize. These GM varieties are generally referred to as Bt varieties, although different companies market them under a variety of trade names. The effect of the use of Bt cotton has perhaps been the most striking. Conventional cotton is very susceptible to insect damage and one-quarter of US insecticide production is used on this one crop. In areas of severe pressure from the three major pests affected by the CryIA protein (tobacco budworm, cotton bollworm and pink bollworm), Bt cotton on average requires 15–20% of the insecticide used on conventional cotton and take-up of Bt varieties in these areas has been very high. In Alabama, for example, Bt varieties accounted for 77% of the cotton planted in the first year that they became available. In areas where Cry1A-controlled pests are less abundant, adoption rates of the new cotton varieties have been low.

The effects of adoption of Bt maize have varied from area to area. The principle maize pest affected by the Cry1A protein is the corn borer and where the corn borer is not particularly abundant, US farmers do not spray against it, preferring to tolerate any losses incurred if an infestation occurs. In these areas, use of Bt maize has led to an increase in yield typically of 10–15%. In areas where the corn borer is abundant, farmers have had to spray heavily and pre-emptively with insecticide to control it. In these areas, use of Bt maize has not increased yields but has led to a decrease in the number of sprays applied per season from an average of 7–8 to an average of 1–2. In areas where the corn borer is not present, the Bt varieties are not used.

The use of Bt crops in the USA is regulated by the Environmental Protection Agency (EPA), and the EPA has been concerned right from the start about the possibility of insects developing resistance to the Bt protein. This would make not only GM Bt crops useless, but also the Bt pesticide used by organic and salad farmers. The EPA insisted that farmers would have to provide refuges of non-Bt crops where insects that developed resistance to Bt would be at a selective disadvantage. This is discussed in more detail in Chapter 5; it has not been popular with farmers but appears to have been successful in preventing resistance from developing.

An unexpected benefit of using Bt maize varieties is that the Bt grain contains lower amounts of fungal toxins (mycotoxins). These compounds, which include potent carcinogens such as aflatoxin and fumicosin, are a particular problem in tropical countries because of the warm and humid conditions that maize grain is stored in. They are associated with a high incidence of throat cancer. In temperate countries they are considered to pose most risk to grain-fed animals and the people most aware of them are horse owners. Nevertheless, there is undoubtedly a low-level presence of these chemicals in the human food chain and the European Commission has set maximum levels for their presence in foodstuffs since 2006. The reduced insect damage in Bt crops means that they are less susceptible to fungal infection. This has been an important selling point in South Africa and adoption rates for Bt maize in that country are over 80%.

The other Cry gene that has been used in plant biotechnology is the CryIIIA gene of *Bacillus thuringiensis* var. *tenebrionsis*. The CryIIIA protein is effective against beetles and potato varieties containing the CryIIIA gene are resistant to infestation by the Colorado beetle, a pest that can devastate potato crops. One such variety, NewLeaf, produced by Monsanto, was on the market in the USA for several years in the 1990s but was withdrawn due to poor sales. US potato plantations are attacked by a number of pests that are not controlled by Bt, in addition to the Colorado beetle, and farmers have turned to new, broad-range insecticides instead of the GM option. The NewLeaf variety also failed to find favour with fast-food chains, a key market for potatoes in North America. Nevertheless, Bt potato may have a role to play in Eastern Europe and Russia where the Colorado beetle is a huge problem.

Bt maize is currently the only GM crop to be grown in significant quantities in Europe (Figure 3.3). European legislation on GM crop planting was tightened in 1999, but Bt corn had already become established in Spain before then and the area of Bt maize cultivation in Spain over recent years has been steady at around 75,000 hectares. This area, of course, is tiny compared with the global area of GM crop cultivation, but it represents the only example of consistent GM crop cultivation in Europe. Glyphosate-tolerant soybean was popular in Romania up to 2006 but its use stopped abruptly when Romania joined the EU in 2007. Bt Maize

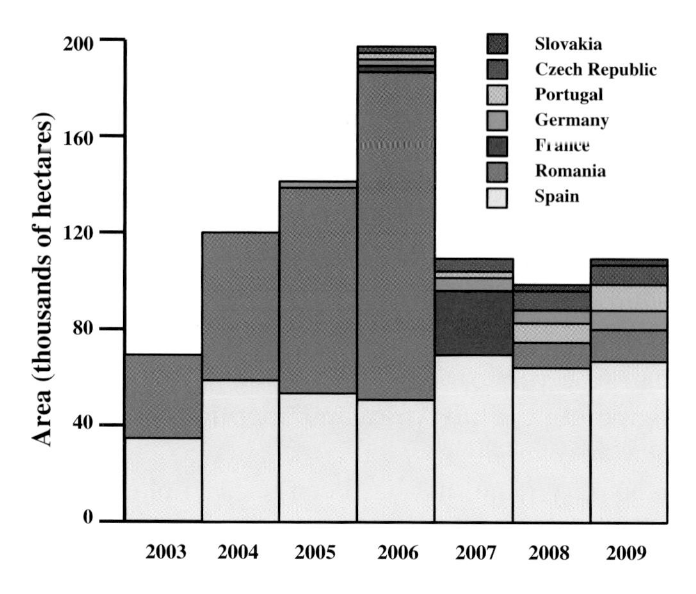

Figure 3.3 GM crop planting in Europe from 2003 to 2009.

appeared to be gaining ground amid great controversy in France in 2006 and 2007; one French farmer was quoted as saying that Bt seed cost him about $48 an acre, instead of about $38 for regular seed, but that he was saving between $24 and $48 an acre because he was not spraying pesticides to kill corn borers. The French government banned cultivation of Bt corn in 2008, despite the fact that the variety had been approved for cultivation throughout the EU (see Chapter 4), putting it at loggerheads with the European Commission.

3.5 Virus Resistance

Plant viruses could be regarded as being as dangerous to humans as viruses causing the worst human diseases. Viruses such as *Cassava mosaic virus* and the *Feathery mottle virus* of sweet potato, for example, are responsible for the deaths of millions of people every year through the destruction of vital food crops. Farmers attempt to prevent viral plant diseases by controlling the insect pests that carry the disease. Those in developed countries also have access to virucidal chemical applications but, as with other agrochemicals, these are expensive and, once established, viral diseases are difficult to bring under control.

The first methods used by biotechnologists to engineer plants to be resistant to viruses arose from investigations into the phenomenon of cross protection, in which infection by a mild strain of a virus induces resistance to subsequent infection by a more virulent strain. Cross protection has been known about for some time and appears to involve the coat protein of the virus because encapsulating a virus in the coat protein of a different virus eradicates it. Engineering a plant to make a viral coat protein was found to mimic cross protection, the plants showing delayed or no symptoms after infection with the virus.

An example of the commercialisation of virus-resistant GM varieties incorporating this technology is a papaya grown in the Puna district of Hawaii. After an epidemic of *Papaya ringspot virus* (PRSV) almost destroyed the industry, growers switched in 1998 to a virus-resistant GM variety. The GM variety contains a gene that encodes a PRSV coat protein. There is no other known solution to an epidemic of PRSV and some commentators claim that the GM variety saved the papaya industry in Hawaii. However, the decision to use the GM option remains controversial even to this day because some markets were lost as a result.

Another method to engineer virus resistance into plants is to use gene suppression techniques (Chapter 2) to block the activity of viral genes when the virus infects. Monsanto, for example, targeted a replicase gene from *Potato leaf roll virus* (PLRV) in this way to induce resistance to PLRV in potato. A potato variety containing this trait and the Bt insect-resistance trait was marketed under the trade name NewLeaf Plus but, like the NewLeaf variety, this was not successful and was withdrawn. Other applications with commercial potential include inducing resistance to *Spotted wilt virus* in tomato.

Virus-resistant papaya, tomato and sweet pepper have been approved for commercial cultivation in China. China has a huge population to feed and over half the population is dependent on agriculture for its livelihood, so the use of plant biotechnology there is extremely important. However, perhaps even more exciting is the potential of this technology in developing countries where losses to viral diseases are the greatest and have the most severe consequences. Kenya, for example, has field-tested a GM sweet potato variety engineered to be resistant to *Feathery mottle virus*. This virus is estimated to reduce Kenya's sweet potato production by half and the yield of the GM variety was reported to be 80% higher than non-GM varieties in the trials. However, these trials were conducted a decade ago and no varieties have yet been made available to farmers. GM cassava that is resistant to *Cassava mosaic virus* may be available sooner. Several teams have addressed this problem, and the Bill and Melinda Gates Foundation and other donors have provided substantial funding to an international consortium of laboratories, enabling real progress to be made. There were reports in 2010 of east African cassava plantations being threatened by another disease, *Brown streak virus*, so the situation is urgent.

3.6 Modified Oil Content

The principal components of plant oils are fatty acids and the various properties of oils from different plants are determined by their differing fatty acid contents. Many hundreds of different fatty acids have been identified in plants. Each fatty acid comprises a carboxylic acid group at the end of a hydrocarbon chain, which is generally regarded as including at least eight carbon atoms (Figure 3.4). Naturally occurring fatty acids are synthesised from the two-carbon molecule, acetyl-CoA, and therefore have an even number of carbon atoms. The fact that they have a non-polar hydrocarbon end terminating in a methyl (CH_3) group and a polar carboxylic acid (COOH) group at the other end means that fatty acids can interact with both water and fats, allowing fats to dissolve in water. Hence they are used in products such as detergents and shampoos as well as foods.

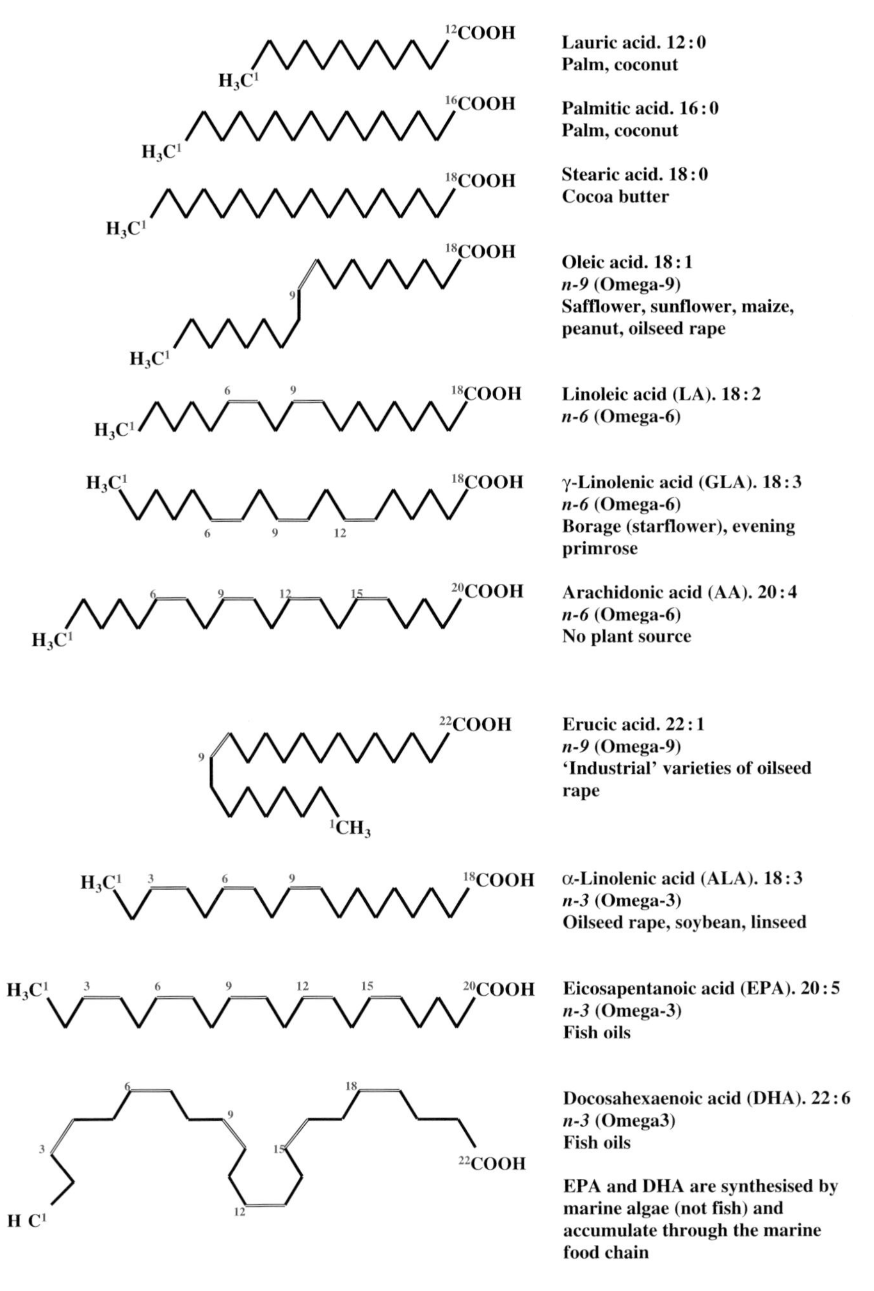

Figure 3.4 Structures of some fatty acids and their sources. Carbon chains are represented by the zig-zag lines, with a carbon atom at each point in the chain.

Figure 3.4 **(*Continued*)** Carbon atoms linked by single bonds to adjacent atoms have two hydrogen atoms attached, while those participating in double bonds have one. Numbers in red indicate the position of a carbon atom in the chain, taking the carbon at the methyl (omega) end as 1, and indicate the first desaturated carbon participating in each double bond.

Fatty acids also have a high energy content and the molecules pack together to form high-density energy stores in the fatty tissues of animals and the seed oils of plants. Furthermore, they are important constituents of cell membranes, while some are converted into hormones and other compounds with vital functions in the human body. As a result, deficiencies in some fatty acids are linked with a range of medical conditions and diseases.

Fatty acids are differentiated not only in chain length but in the number and position of double bonds between the carbons in the chain. Carbon atoms joined by double bonds are described as unsaturated, while those that are joined by single bonds are saturated. Fatty acids in which all of the carbons are joined by single bonds are similarly called saturated fats, while those with a single double bond are monounsaturates and those with more than one are polyunsaturates. Long-chain saturated fats are the major components of animal fat stores, while plant oils contain a variety of fatty acids with different chain lengths and degrees of saturation (Table 3.2).

The number of carbon atoms and double bonds in the chain is given as a ratio; oleic acid (Figure 3.4), for example, is an 18-carbon fatty acid with a single double bond (a monounsaturate), so the ratio is 18:1. The position of the first double bond in the chain is given in the form *n-x*, where *x* is the position of the first unsaturated carbon with respect to the methyl (omega) end of the molecule; so, for oleic acid, this number is *n-9*. Oleic acid (18:1, *n-9*) is also known as an omega-9 fatty acid for this reason.

Well-known plant fatty acids (Figure 3.4) include lauric acid (12:0) and palmitic acid (16:0), which are the prevalent fatty acids in coconut and palm kernel oil. Lauric acid is used in the

Table 3.2 Typical fatty acid content (%) of some plant oils. Where total does not equal 100%, balance is made up of minor constituents that are not shown. OR = oilseed rape.

	Caprylic acid 8:0	Capric acid 10:0	Lauric acid 12:0	Myristic acid 14:0	Palmitic acid 16:0	Stearic acid 18:0	Oleic acid 18:1 Omega-9	Linoleic acid; LA 18:2 Omega-6	α-Linolenic acid; ALA 18:3 Omega-3	γ-Linolenic acid; GLA 18:3 Omega-6	Erucic acid 22:1 Omega-9
Oilseed rape (Edible)					4	2	61	21	11		
High lauric GM OR			40		3	1	40	15			
Oilseed rape (Industrial)				1	3	1	14	11	11		54
Sunflower					9	7	10	74			
Palm				1	44	5	39				
Soybean					11	4	23	54	7		
Plenish GM soybean					11	4	80				
Coconut	8	7	48	18	9	3	6				
Linseed					6	4	20	16	54		
Evening primrose					6	2	8	70		10	
Starflower					10	4	8	37		23	
Maize					12	2	30	54	1		
Peanut						18	47	29			
Olive					14	2.5	69	12			

manufacture of cosmetics and detergents: for example, the sodium lauryl sulphate that is commonly used in shampoos is derived from it. Palmitic acid is used in both detergents and foods. Stearic acid (18 : 0) is a major component of cocoa butter, while oleic acid (18 : 1, *n-9*) is the major constituent of olive and oilseed rape oil. Linoleic acid (LA) (18 : 2, *n-6*) is found in safflower, sunflower and maize oil, and makes up about 20% of oilseed rape oil. Gamma linolenic acid (GLA) is another omega-6 fatty acid and is identical to LA except that it has an additional double bond at *n-12*; starflower (borage) oil contains more GLA than any other but evening primrose oil is also a good source and in the UK at least is generally much cheaper.

LA is an essential fatty acid (meaning an essential part of our diet) because the human body requires it but cannot make it. GLA is also often described as essential because, although the human body can make GLA, it requires LA to do it; it cannot make GLA *de novo*. Lack of these fatty acids causes dry hair, hair loss and poor wound healing. One of their uses is to be converted to arachadonic acid (AA; 20 : 4, *n-6*) (Figure 3.4), with GLA being an intermediate between LA and AA (this is significant because most people consume more LA than GLA). AA is present in the phospholipids of cell membranes and is abundant in the brain and muscles. Long-chain polyunsaturated fatty acids are also used to make a family of compounds called eicosanoids, which include prostaglandins, leukotrienes and isoprostanes. These compounds perform a number of essential physiological functions including regulation of the immune system, blood clotting, neurotransmission and regulation of cholesterol metabolism. Dietary intake of GLA may be particularly important for people with diabetes, who appear to convert LA to GLA inefficiently. AA itself is found in meat, eggs and dairy products but there is none whatsoever in a vegan diet. It is therefore particularly important for vegans that they consume adequate amounts of LA or GLA from plant sources. Many animals, including the domestic cat and other felines, are unable to synthesise AA from LA or GLA at all and consequently are obligate carnivores.

Erucic acid (22 : 0) is another interesting plant fatty acid. It is poisonous but has a variety of industrial uses. It has similar properties to mineral oils, but is readily biodegradable, and is used in transmission oils, oil paints, photographic film and paper emulsions, healthcare products, plastics (in the form of its derivative, erucamide) and biodiesel. Erucic acid used to make up about 50% of oilseed rape oil, making the oil unfit for human consumption. Oilseed rape also had the problem that the meal left over after oil extraction contained high levels of compounds called glucosinolates, which are also poisonous and have a bitter taste. Oilseed rape was used as a forage crop in the UK in the nineteenth century but the first time that it was grown widely in the United Kingdom was during the Second World War, when it was used to produce industrial oil. Plant breeding of different varieties over the next thirty years reduced the levels of erucic acid and glucosinolates to the point where these varieties were considered acceptable for human consumption. This process involved an intense programme of mutagenesis.

The first low erucic acid, low glucosinolate varieties were grown in Canada in 1968. Nevertheless, oilseed rape did not get its seal of approval for human consumption (Generally Recognized as Safe) from the Food and Drug Administration of the USA until 1985. Canadian producers then came up with the name Canola (Canadian oil low erucic acid) for edible oilseed rape oil. This name was adopted all over North America not only for the edible oil but also for the crop itself.

The oil of modern oilseed rape varieties for human and animal consumption is made up of oleic acid (60%), LA (20%) and alpha linolenic acid (ALA) (10%), with palmitic, stearic and other fatty acids together accounting for the other 10%. High erucic acid varieties are still grown today for industrial purposes but are not permitted to be placed in the food chain. The rapidly increasing demand for biodiesel may lead to more cultivation of high erucic acid varieties and conflicting demands on plant breeders and growers for varieties with different oil profiles suitable for different end uses.

Oilseed rape oil has already been the target for biotechnologists, partly because the oil is one of the cheapest edible oils on the market, so growers and processors are always likely to be interested in anything that adds value to it. The company Calgene, subsequently taken over by Monsanto, genetically modified an oilseed rape variety to produce high levels of lauric acid in its oil. This variety was introduced onto the market in 1995. It contains a gene from the Californian Bay plant that encodes an enzyme that causes premature chain-termination of growing fatty acid chains. The result is an accumulation of the 12-carbon chain lauric acid to approximately 40% of the total oil content, compared with 0.1% in unmodified oilseed rape, making the oil an alternative to palm and coconut oil as a source of lauric acid for detergents. However, palm oil production in particular is very efficient and oil from the GM oilseed rape did not gain a foothold in the market. Cultivation of this variety was never more than small-scale and has now ceased.

The other major crop that has been genetically modified to increase the value of its oil is soybean. The genetically modified variety was produced by PBI, a subsidiary of DuPont. It accumulates oleic acid to approximately 80% of its total oil content, compared with approximately 20% in non-GM varieties. In conventional soybean, much of the oleic acid is converted to LA by an enzyme called a delta-12 desaturase and some of the LA is further desaturated by a delta-15 desaturase to ALA (Figure 3.4). In the GM variety, the activity of the gene that encodes the delta-12 desaturase is reduced so that oleic acid levels are increased while LA and ALA levels are decreased. Note that, somewhat confusingly, desaturases are usually named according to the position of the first carbon in the double bond that they introduce with respect to the acidic or delta end of the molecule, rather than the omega end.

LA and ALA are essential amino acids and are nutritionally important. However, oleic acid is very stable during frying and cooking and is less prone to oxidation than polyunsaturated fats, making it less likely to form compounds that affect flavour. The traditional method of preventing polyunsaturated fat oxidation involves hydrogenation and this runs the risk of creating *trans* fatty acids. *Trans* fatty acids contain double bonds in a different orientation to the *cis* fatty acids present in natural plant oils. They behave like saturated fat in raising blood cholesterol and potentially contributing to blockage of arteries. The oil produced by high oleic acid GM soybean requires less hydrogenation and there is less risk of *trans* fatty acid formation.

As with the high lauric acid oilseed rape, the advantage to farmers of growing high oleic acid soybean is that they get a premium price for it. The variety is marketed under the trade name Plenish. Monsanto also have a GM low-linolenic variety called Vistive, but the modified fatty acid trait was developed conventionally. Initial take-up of low-linolenic varieties was not impressive but it increased when the US government introduced legislation requiring that information on *trans* fatty acid content be included in food labels.

Reducing *trans* fatty acid content of foods is a worthy target but conventional soybean is also a valuable source of LA and ALA. It is important, therefore, that consumers are made aware of the changes in fatty acid content, with possible disadvantages made clear. Linseed oil is an alternative source of ALA. However, it has a strong flavour and 'industrial' odour, so it is not consumed in large quantities, although it is taken by some people as a nutritional supplement.

The examples above are undoubtedly only the first of many GM crops modified to change their oil content. Many oils have uses in the plastics and industrial oil industries and there are examples of crops that have been modified to make oils of this type. Increasingly, plant oils are also now being used to make biodiesel, usually after the fatty acids have been esterified with methanol to create fatty acid methyl esters (FAMEs). Biodiesel production is now a major and rapidly growing industry in Europe.

Another exciting prospect is the production of oils with nutritional or pharmaceutical (sometimes called nutraceutical) properties. As discussed above, animals, including humans, have only a limited capacity to synthesise some fatty acids and a number of fatty acids, including LA and ALA, that are produced by plants, have been identified as essential dietary components (essential fatty acids, or EFAs).

In the case of GLA, which can only be made relatively inefficiently from LA, the plant species for which it is a major oil constituent, namely evening primrose and starflower (borage), make poor crops. An early target for biotechnologists, therefore, was to take a gene from starflower that encodes a delta-5 desaturase and engineer it into oilseed rape or sunflower, both of which accumulate LA, to convert some of the LA into GLA. This has been achieved in the model plant, Arabidopsis, but the technology has not been commercialised because, despite the problems in sourcing GLA from evening primrose and starflower, the cost of the oil itself makes up a relatively small part of the total cost of the product by the time it reaches the consumer. There is, therefore, no economic incentive to produce the oil more cheaply. However, further engineering of Arabidopsis by a team led by Johnathan Napier at Rothamsted Research and the University of Bristol in the UK has shown that it is possible to produce GM plants that accumulate AA itself.

Another application being pursued by Johnathan Napier's team and by several biotech companies is the production of marine fish oils in plants. Marine fish oil contains long chain, omega-3, polyunsaturated fatty acids (LC-PUFAs) such as eicosapentaenoic acid (EPA) (20:5) and docosahexaenoic acid (DHA) (22:6) (Figure 3.4). There is no plant source of these omega-3 LC-PUFAs. The human body can synthesise them from ALA but the efficiency is low: approximately 5% in men and slightly higher in women.

As with the omega-6 LC-PUFAs such as AA, the omega-3 LC-PUFAs are precursors for eicosanoids but the eicosanoids derived from omega-3 and omega-6 LC-PUFAs have distinct properties. In general, eicosanoids derived from omega-6 LC-PUFAs are pro-inflammatory, pro-aggregatory and immuno-active, while those derived from omega-3 LC-PUFAs have little or no inflammatory activity and act to modulate platelet aggregation and immuno-reactivity.

Foetal and infant development is now known to require omega-3 LC-PUFAs and both EPA and DHA are commonly added to infant formula milk. There is also increasing evidence of the effectiveness of omega-3 LC-PUFAs in the prevention of cardiovascular disease. This link was first proposed in the 1970s as a result of studies on Inuit populations who consumed fat-rich diets but suffered almost no cardiovascular disease. The fish oils consumed by these people were rich in omega-3 LC-PUFAs, and follow-up studies supported the hypothesis that it was these compounds that were protecting the Inuit from the disease.

More recently, it has also emerged that omega-3 LC-PUFAs can protect against metabolic syndrome, a collection of pathologies that are indicative of progression towards cardiovascular disease, obesity and type 2 diabetes. Treatment for metabolic syndrome now includes dietary intervention with reduced carbohydrate intake and increased fish oil consumption.

The problem with marine fish as a source of omega-3 LC-PUFAs such as EPA and DHA is that marine fisheries are in a global crisis with stocks diminishing through over-fishing. There are also concerns about the levels of environmental pollutants such as dioxins, PCBs and heavy metals in fish oils. For example, the use of fish oils in products for babies and young children is not permitted in the USA. Furthermore, fish are no more capable of synthesising these compounds *de novo* than other animals are. In fact, omega-3 LC PUFAs are synthesised by marine algae and accumulate through the marine food chain; in other words fish acquire them in their diet. This means that farmed fish are not a sustainable source of these essential dietary components; they only accumulate them because they are fed fishmeal from marine fish. Finding a sustainable alternative source is therefore essential and GM plants producing EPA and DHA are being developed.

Engineering plants to produce these very long chain PUFAs from ALA has not been a simple task because of the numbers of genes involved (Figure 3.5). However, the teams at Rothamsted Research and the University of Bristol have succeeded in engineering the model plant, Arabidopsis, to produce EPA. Monsanto, BASF and DuPont are now all believed to be developing commercial varieties of soybean and oilseed rape that produce EPA and DHA.

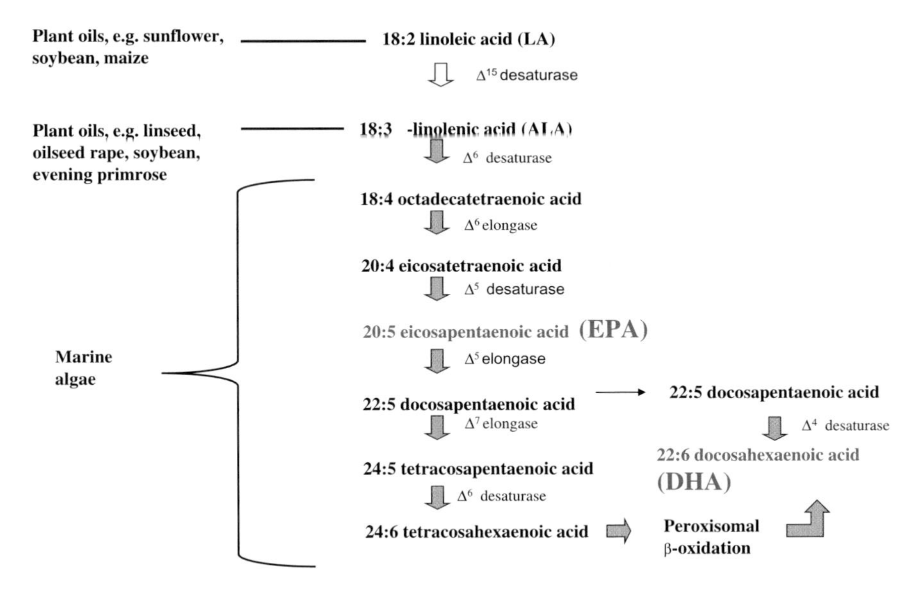

Figure 3.5 Pathways for biosynthesis of very long chain omega-3 polyunsaturated fatty acids (omega-3 LC PUFAs) that are present in fish oils, including eicosapentanoic acid and docosahexaenoic acid. These PUFAs are actually synthesised by marine algae but accumulate through the marine food chain. The longest omega-3 PUFA in crop plants is α-linolenic acid (ALA).

3.7 Modified Starch for Industrial and Biofuel Uses

Starch from cereals, potato, sweet potato and cassava is the most important carbohydrate in the diets of humans and farm animals and by far the most important dietary energy source. Over two billion tonnes of starch are produced annually in cereal grain and seven hundred million tonnes in roots and tubers. In developed countries, starch provides 35% of daily calorific intake, while in developing countries it may provide 80%. Starch is also used as a thickener in the food industry, while non-food uses include the manufacturing of paper (starch typically makes up 8% of a sheet of paper), adhesives, gypsum wall boards and textile yarns.

A difficulty in using starch for industrial purposes or as a food additive is that the two components, amylose and amylopectin, have different characteristics and have to be separated or modified chemically before use. Amylose, for example, has gelling properties that are undesirable in some processes. The relative amounts of amylose and amylopectin are determined by the activities of different starch synthases and BASF has developed a genetically modified potato, marketed as Amflora, in which the activity of granule-bound starch synthase is greatly reduced. The starch from this new potato variety is composed almost entirely of amylopectin, whereas normal potato starch contains approximately 20% amylose. Amflora was developed for the European market and was mired in the European Union's tortuous regulatory process covering the use of genetically modified crops for over a decade. It was finally approved for cultivation in 2010. BASF has announced its intention to cultivate the Amflora variety in Germany and Sweden in 2011. The area of cultivation is tiny and the intention is presumably to produce seed potatoes for more widespread cultivation in subsequent years.

Starch can be used to produce sugars through enzymatic digestion. Until recently these sugars were used largely in the food industry, but over the last few years there has been a huge increase in the use of sugars derived from starch for the production of ethanol for fuel. This industry is now well established in the USA, where the annual growth rate in ethanol production, almost entirely from maize starch, was 25% between 2003 and 2007. In 2010, bioethanol production took a third of the US maize crop, a staggering figure. The UK is not far behind. A plant designed to produce biofuel from wheat grain has been built on Teeside; it has suffered from teething troubles but when fully operational it is expected to take 1.2 million tonnes of grain per year (about 8% of the total UK wheat crop in a good year), and produce 400 million litres of ethanol, 350 thousand tonnes of animal feed and 300 thousand tonnes of carbon dioxide for use in the manufacture of soft drinks. A similarly sized plant will come on line in the UK in 2011, operated by Vivergo, a joint venture between British Sugar, BP and DuPont. With a number of smaller plants in the pipeline, one fifth of the UK's wheat harvest could be used for fuel rather than food production by 2015.

In principle, bioethanol production from cereal grains is no different to the production of alcoholic drinks by malting, fermenting and distilling. During traditional malting, cereal grains are exposed to moist, warm conditions that induce partial germination and allow the grains' own starch-degrading enzymes to be synthesised. This has been considered too complex for bioethanol production for fuel and a 'dry-grind' process has been favoured in which the entire kernel is ground into a coarse flour, then slurried with water. The resulting mash is then cooked, treated with enzymes, fermented and distilled. The first enzyme to be added is α-amylase, an endoamylase that acts at random locations along the starch chain to yield shorter glucan chains. Ultimately its products are maltotriose, maltose and limit dextrin (a mixture of branched and unbranched glucans) and the process is known as gelatinisation and liquefaction. Gluco-amylases are then added to produce smaller sugars that can be fermented; this process is known as saccharification.

Syngenta have produced a GM maize variety with a highly thermostable α-amylase gene, *amy797E*, from the thermophilic bacterium, *Thermococcales*. This variety is claimed to give a better yield of ethanol in the dry-grind process. It has just been deregulated by the US authorities (early 2011).

The demand for starch and starch-containing crops has been growing more rapidly than starch production for many years, culminating in stocks reaching alarmingly low levels in 2005–2006 and a consequent steep rise in prices. The situation was exacerbated by a long-term drought in Australia, one of the major wheat exporters. Farmers responded to the rise in prices by increasing production, and prices fell back again. However, 2010 saw another sharp increase in prices, this time as a result of a severe drought in Russia, another major wheat exporter, which culminated in the Russian government banning wheat exports in order to ensure that they had sufficient grain to meet domestic demand. The long-term trend in demand continues upwards as more grain is used to produce bioethanol, world population expands and people in emerging countries want to eat more, better food and more meat (which means more grain for animal feed) and have the money to pay for it. Meeting that demand in the coming years will be a challenge.

3.8 High Lysine Corn

The use of grain starch for bioethanol production leaves a high-protein co-product that is an ideal raw material for animal feed. An example of a variety that has been produced with both co-products in mind is Mavera maize, which was developed by Renessen, a joint venture between Cargill and Monsanto. Mavera combines a triple stack of input traits: resistance to corn rootworm and the European corn borer and tolerance of glyphosate. The novelty, however, is that the grain has a relatively high lysine content. Lysine is an essential amino acid and cereal grains are generally too low in it to provide a balanced diet on their own. This means that animal feed must contain other sources of lysine, for example soybean or oilseed rape meal. Indeed, this is one of the reasons why home-grown barley has lost most of the UK animal feed market to imported soybean.

The high lysine trait is imparted by a bacterial gene encoding a lysine-insensitive dihydrodipicolinate synthase. Dihydrodipicolinate synthase is an enzyme in the pathway for lysine synthesis which is feedback-inhibited by lysine, thereby providing the major regulatory control for flux through the pathway. The bacterial enzyme is not affected by lysine, allowing the amino acid to accumulate to levels beyond what it would in an unmodified plant.

The benefits of Mavera maize are reduced costs and a more nutritious feed that does not require supplementation in order to provide animals with adequate amounts of lysine. Currently it is being grown entirely for US domestic bioethanol production with high-lysine animal feed as a valuable co-product. However, it should sound alarm bells for European maize growers who may find themselves competing with varieties like Mavera but unable to grow them themselves because of the stifling effect of EU regulations. This is discussed in more detail in the following chapters.

3.9 Vitamin Content: Golden Rice

Vitamins are compounds that are required for the normal functioning of our bodies but which we cannot make ourselves. They are, therefore, an essential component of our diet. They are involved in physiological processes such as the regulation of metabolism, the production of energy from fats and carbohydrates, the formation of bones and tissues and many others.

While nutritionists still argue over the optimum levels of different vitamins that people should consume, it is agreed that consumers in the developed world have access to foods (fresh fruit, vegetables, bread, meat, dairy products) containing all of the vitamins that they require. They also have access to an array of vitamin supplements. However, while most people eat plenty of meat and dairy products, a significant number avoid them altogether, and few people eat the recommended amounts of fresh fruit and vegetables. The health of some people might, therefore, be improved if the levels of vitamins in foods were increased. In other words, if vitamin levels were increased in foods that consumers like instead of consumers being persuaded to change their diets. Furthermore, for foods that consumers associate with a healthy lifestyle, such as breakfast cereals, there is market advantage to be gained in improving the nutritional value of the product and advertising the fact to consumers.

Obviously, manufacturers have the option of adding vitamins to their products but in some cases this is expensive. This might present opportunities for biotechnologists to engineer crop plants to contain more vitamins. Folic acid, deficiency of which may cause gastrointestinal disorders, anaemia and birth defects, is one possible target. Others are the fat-soluble vitamins E and K, deficiencies in which are associated with arterial disease and, in the case of vitamin K, post-menopausal osteoporosis.

The more obvious need for increasing the vitamin content of peoples' diets is in the developing world, where acute vitamin deficiency is the cause of poor development, disease and death. Vitamin A deficiency, for example, is common in children in developing countries who rely on rice as a staple food. It causes symptoms ranging from night blindness to those of xerophthalmia and keratomalacia. These related conditions lead to pathologic dryness of the conjunctiva and cornea. The conjunctiva may become thick and wrinkled and the cornea ulcerated, leading to total blindness. Vitamin A deficiency is the leading cause of blindness in children in the developing world: over 5 million children develop xerophthalmia annually and approximately 250,000 of these become blind. Vitamin A deficiency also exacerbates diarrhoea, respiratory diseases and measles; improving vitamin A status in children reduces death rates by 30–50%. Efforts to eradicate vitamin A deficiency are co-ordinated by the World Health Organisation but have failed so far due to the difficulty of reaching those in need.

Ingo Potrykus, a biotechnologist at the Swiss Federal Institute of Technology in Zurich, saw the possibility of using a complementary approach through the genetic modification of rice, the staple crop for many subsistence farmers. Rice grain does contain vitamin A but only in the husk. The husk is discarded because it rapidly goes rancid during storage, especially in tropical countries. Potrykus aimed to produce a rice variety that would make pro-vitamin A (beta-carotene, a precursor that humans can process into vitamin A) in its seed endosperm, the largest part of the seed that is eaten. A conventional breeding approach would not work because no conventional rice variety makes any pro-vitamin A at all in its endosperm.

Rice endosperm synthesises a compound called geranylgeranyl diphosphate, which is an early intermediate in the pathway for beta-carotene production. Potrykus' team successfully engineered the rest of the pathway (Figure 3.6) into rice, using phytoene synthase (*psy*) and lycopene β-cyclase genes from daffodil (*Narcissus pseudonarcissus*), and a phytoene desaturase (*crtI*) gene from the bacterium *Erwinia uredovora*.

The GM rice producing pro-vitamin A was crossed with another line containing high levels of available iron. Rice normally contains a molecule called phytate that ties up 95% of the iron,

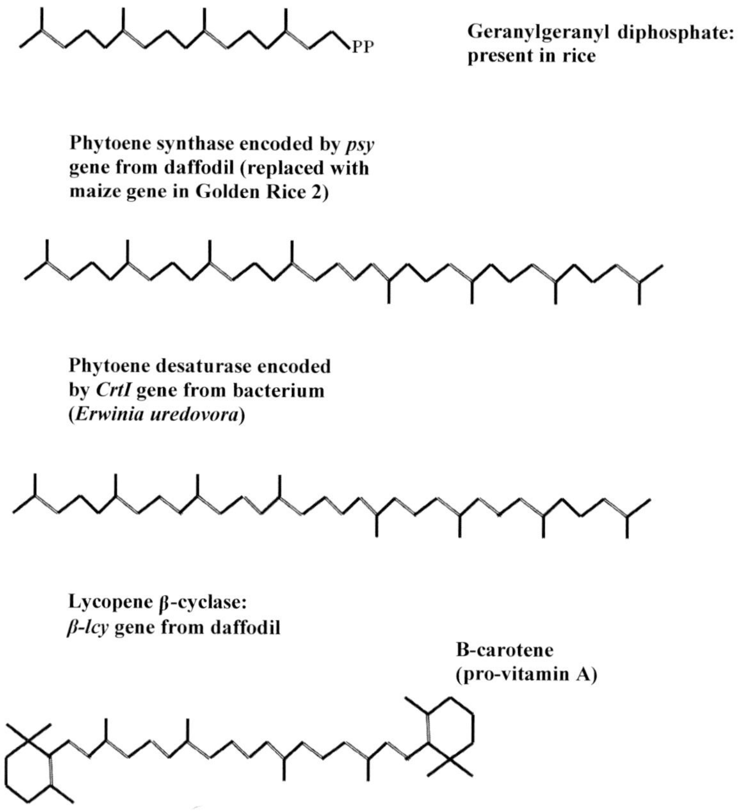

PP

Geranylgeranyl diphosphate: present in rice

Phytoene synthase encoded by *psy* gene from daffodil (replaced with maize gene in Golden Rice 2)

Phytoene desaturase encoded by *CrtI* gene from bacterium (*Erwinia uredovora*)

Lycopene *β*-cyclase: *β-lcy* gene from daffodil

Β-carotene (pro-vitamin A)

Figure 3.6 Engineering of the pathway for β-carotene (pro-vitamin A) synthesis from geranylgeranyl diphosphate in Golden Rice.

preventing its absorption in the gut. The GM rice contains a gene encoding an enzyme called phytase that breaks phytate down. The high pro-vitamin A/high available iron hybrid was called Golden Rice.

The aim of the Potrykus team was to enable people to obtain sufficient pro-vitamin A to avoid the symptoms of vitamin A deficiency in a daily ration of 300 g of rice. Anti-GM pressure groups, with Greenpeace at the fore, claimed that much larger quantities of the rice would have to be eaten to achieve the Recommended Daily Allowance (RDA) level. In their press release of February 2001, for example, Greenpeace described Golden Rice as 'fool's gold', claiming that an adult would have to eat at least 3.7 kg of dry rice (12 times the normal intake of 300 g) to get the daily recommended amount of pro-vitamin A. The press release coincided with the start of a breeding programme undertaken by the International Rice Research Institute in the Philippines to cross Golden Rice with local rice varieties. Potrykus replied that the calculations of Greenpeace were based on RDA values for western consumers, who have access to as much vitamin A as they care to eat, and that the levels of pro-vitamin A achieved with Golden Rice were in the 20–40% range of the RDA in a 300 g serving. Nutritional experts involved in the project believed that this level would have a significant effect in preventing blindness and other symptoms associated with severe vitamin A deficiency. Even so, the full RDA would be provided by 0.75–1.5 kg dry weight, not 3.7 kg. The debate fizzled out anyway when Golden Rice 2 was announced. Golden Rice 2 was developed by Syngenta in collaboration with Potrykus by replacing the daffodil phytoene synthase gene with one from maize and it contains many times more pro-vitamin A than Golden Rice 1.

The production of Golden Rice was first reported in 2000. At that time it was an experimental breeding line and it had to be crossed into local varieties around the world to produce varieties that could be made available to farmers. That process was led by centres such as the Rice Research Institute in Manila, the Philippines. However, governments are very protective of their export markets in Europe and Japan, where resistance to GM crops has been strong, and have also been alarmed by the anti-GM stance taken by some countries. This has undoubtedly held up the release of varieties to farmers and that now seems unlikely to happen before 2012 or even 2013. The process has taken perhaps ten years longer than it needed to because of what might be called 'European attitudes' to GM crops. This may come back to haunt European regulators and pressure groups if high pro-vitamin A rice is seen to save lives.

Rice is not the only crop in which this technology is potentially applicable and a programme to engineer pro-vitamin A synthesis into cassava, for example, is reported to be well advanced.

3.10 Fungal Resistance

Fungal diseases of plants cause severe losses in crop production. An extreme example is the Irish potato famine of the nineteenth century, which was caused by a pandemic of the fungus *Phytophthora infestans*, which causes late blight disease. Resistance to infection on the part of the plant is imparted in many cases by so-called resistance or R genes. R genes encode proteins that act as receptors for pathogen molecules and induce a hypersensitive response (HR) in which there is rapid cell death around the entry point of the fungus, preventing development of the disease.

The pathogen proteins that are detected by R genes are encoded by avirulence genes (*Avr* genes). An *Avr* gene aids infection of plants that do not carry an R gene that recognises it, but must be discarded by the pathogen to overcome resistance imparted by an R gene. Thus there is a complicated relationship between the genetics of plants and their fungal pathogens that has arisen through millions of years of co-evolution.

The success of R genes is dependent on there not being too many individuals carrying any one particular R gene. There is then some selective advantage in a pathogen retaining an *Avr* gene, in that it is better able to infect those plants without the R gene. In a field of crop plants that are essentially all the same this selective advantage does not exist. As a result, the efforts of plant breeders to breed resistance into crop varieties by introducing particular R genes have rapidly been overcome by the emergence of new strains of the pathogen that are not affected. Nevertheless, biotechnologists believe that if they can learn more about how the R gene system works they will be able to engineer fungal resistance into crop plants, perhaps by stacking multiple R genes in one crop variety.

Biotechnologists have also attempted to engineer resistance to late blight into potato by finding wild species of potato that are resistant to the disease, identifying the genes that impart resistance and transferring them to cultivated potato varieties by genetic modification. In theory, moving genes from wild potato species into breeding programmes is possible with conventional breeding methods. However, wild varieties are diploid, while cultivated varieties are tetraploid (in other words there has been a doubling of the chromosome number in the development of cultivated potato). This makes crossing wild and cultivated potato species difficult. Furthermore, potatoes contain toxic compounds known as glycoalkaloids, including solanine and chacocine, which deter insects and other herbivores and have antifungal properties. These large organic compounds (the molecular formula for solanine is $C_{45}H_{73}NO_{15}$) cause nausea, dizziness, vomiting, diarrhoea, heart arrhythmia and in extreme cases coma and death. Their levels in the tubers of cultivated potatoes have been reduced by breeders and in the UK must comply with a regulatory limit of 200 mg per kg (although levels increase if the tubers are exposed to light and start to green). The tubers of wild potato species usually contain much higher levels and are not considered fit for human consumption. If a cross is made between a wild and cultivated species it requires a lengthy process of back-crossing to reduce the glycoalkaloid concentration back to acceptable levels. Moving genes from wild to cultivated potato species is therefore an example of where GM is clearly the most efficient and safest route.

BASF have used a gene called *RB* from a wild potato species, *Solanum bulbocastanum*, to confer broad-spectrum resistance to late blight in a cultivated potato variety. The company initially chose Ireland to field-trial the GM potato lines in 2006, but this met with fierce controversy and the plans were abandoned. In 2007, two trials were undertaken in the UK, but the trial sites were vandalised and it is not clear how much data the company were able to obtain.

In 2010, two GM potato lines were field-trialled at the John Innes Centre in Norwich. One contained the *Rpi-vnt1.1* gene from *Solanum venturii*, the other the *Rpi-mcq1* gene from *Solanum mochiquense*. Both GM lines had shown resistance to late blight in glasshouse experiments. In the field, resistance in the line carrying the *Rpi-mcq1* gene broke down, but the line carrying the *Rpi-vnt1.1* gene performed well. This was only one field trial and one season, and late blight is clearly an extremely difficult nut to crack, but this result is encouraging.

An alternative approach to the challenge of engineering fungal resistance into crop plants is to modify them with genes that express fungicidal proteins. Examples are genes encoding the enzymes chitinase and β glucanase, both of which attack the cell walls of fungal hyphae as they enter the plant. Transgenic plants containing genes for these enzymes have been reported to have increased resistance to pathogenic fungi under experimental conditions.

3.11 Drought, Heat and Cold Tolerance; Climate Change

The availability of water is a major determinant of plant production in many parts of the world, and shortages of water are recognised as major threats to food security. There is therefore an urgent need for new, high-yielding varieties with enhanced drought tolerance. However, developing high-yielding crops for water-limited environments is a major challenge and progress has been limited by the complexity of the underlying traits, which are often determined by multiple genes and the seasonal and year-on-year variation of water availability. Drought is often, although not always, accompanied by high temperatures and although these two stresses provoke different responses in plants the dividing line is often blurred as one stress exacerbates the effects of the other.

While both drought and heat stress are already problems in many parts of the world, the range over which they impact seriously on crop yields and the frequency with which they do it are both predicted to increase as a result of climate change. If the predictions of climate change are correct, plant breeders face the challenge of developing crop varieties for an environment that is going to change over the coming decades at unprecedented speed. The upper end of the range predicted by Global Climate Models in the Intergovernmental Panel on Climate Change (IPCC) 4th Assessment Report is an increase of 5 °C in global mean temperature by the end of the century; this is a rate of change the like of which has not occurred in the last 50 million years. As well as the mean increase in temperature, there is predicted to be an increase in frequency of extreme weather events.

The potential impact of such events on food production can be seen in the effects of the severe Australian drought of the last decade, which reached crisis point in 2007–2008, and the Russian drought of 2010; both of these events were alluded to in Section 3.7 on starch. Both Australia and Russia are major wheat exporters in normal years, but in 2007–2008 exports of wheat from Australia fell to 1.7 million tonnes, compared with 7.5 million tonnes in 2005–2006, while in July 2010 the Russian government banned wheat exports altogether to protect domestic supply. The price of wheat grain on the London International Financial Futures & Options Exchange (LIFFE) rose in 2008 to £198 per tonne. It fell back to below £100 per tonne in 2009 but rose to a new record of £200 per tonne in December 2010.

In both cases, these droughts were accompanied by high temperature and both stresses would probably have affected grain yield. Drought stress can be devastating at any time during wheat development, while increased temperature shortens the growing period, reducing yield. Heat stress at flowering is especially damaging, resulting in much lower grain number and substantial yield losses. During the hot summer in northern Europe in 2003, for example, with maximum temperatures up to 38°C in the UK and 40°C in France, wheat production fell by about 20%.

Plants adopt several strategies to avoid the effects of drought. For example, if drought is most likely to occur in late summer (typical of the UK and northern Europe) they may avoid it by growing, flowering and setting seed before this time. Other 'avoidance' strategies include the development of deeper and more extensive root systems, allowing the plant to obtain more water from the soil to survive dry periods. Plants have also evolved responses that enable them to survive even if they do become short of water. These are referred to as tolerance traits and they differ from species to species and between different varieties, developmental stages, organs and tissue types. The plant hormone, abscisic acid (ABA), plays a key role, initiating a network of signalling pathways involving multiple protein kinases (enzymes that attach a phosphate group to another protein, affecting its activity) and transcription factors (proteins that regulate the expression of genes).

Transcription factors that are known to be involved in water stress responses include dehydration-responsive element binding protein (DREB)-1 and -2, ABA response element binding proteins (AREBPs), members of the zinc finger homeodomain (ZFHD)-1, myeloblastosis (MYB) and myelocytomatosis (MYC) families and the NAC family, which comprises no apical meristem (NAM), ATAF1 and 2, and cup-shaped cotyledon (CUC) transcription factors (these apparently odd names derive from the characteristics of mutant plants in which the transcription factor is not present). The action of AREBPs, DREB1, MYC and MYB requires ABA, while that of DREB2, ZFHD1 and NAC is ABA-independent. Over-expression of another transcription factor, plant nuclear factor-Y (NF-Y) has been shown to confer increased drought tolerance in maize in the field.

One of the areas of plant metabolism that is affected by these signalling pathways is carbohydrate metabolism, which plants manipulate to mitigate the effects of osmotic stress brought about by drought, for example by inter-converting insoluble starch with soluble sugars. Many of these responses are also seen in plants in response to cold, because cold, like drought, causes plant cells to suffer osmotic stress. Over-expression of protein kinases involved in ABA signalling has been shown to improve cold tolerance under laboratory conditions. Freezing is particularly dangerous to a plant because ice formation outside a plant cell prevents water uptake and causes an extreme osmotic stress leading to malfunctioning of cellular membranes, cellular dehydration and irrevocable damage. Plants prevent frost damage by modulating the concentrations of solutes and the lipid composition of their membranes. In 1991, a company called DNA Plant Technology attempted to improve the frost tolerance of tomato by introducing a gene encoding a so-called 'antifreeze' protein from a fish, the winter flounder. The project did not get beyond limited field trials and although similar experiments with other crop species have been performed it seems unlikely that such crops will reach the market in the near future.

The prospect of crop plants being engineered with animal genes was a particularly emotive issue and the 'fishy tomato' became an icon of the anti-GM campaign (see Chapter 5).

Many of the genes that are expressed in response to drought and cold stress are also expressed when heat stress is applied. In some cases this may be misleading because elevated temperatures will cause water stress unless ambient humidity is adjusted to prevent it, and this sometimes obscures the fact that heat stress presents a plant with its own specific problems. As described above, high temperature causes wheat and other cereals to develop and mature more quickly; it also brings about an increase in respiration and an inhibition of photosynthesis. The latter is caused by a reduction in the activity of ribulose 1,5-bisphosphate carboxylase/oxygenase (Rubisco) and the efficiency of photosystem II. The reduction in Rubisco activity occurs because the enzyme responsible for maintaining its activity, Rubisco activase, is unstable at even moderately high temperatures. Genetic manipulation of Rubisco activase to improve its stability at high temperatures is therefore a potentially important target.

Another effect of high temperature is oxidative damage and high activities of superoxide dismutase and catalase have been shown to be associated with good thermo-tolerance in wheat. Temperatures greater than 35°C during wheat grain development have also been shown to cause changes in the expression of different groups of seed storage proteins, with consequent effects on dough quality.

Transcription factors that are specifically associated with heat stress include heat shock factors (HSFs). The HSFs are quite a large family with at least 21 members, and their interaction appears to be complex. Nevertheless, simple over-expression experiments with HSFs have resulted in increased thermo-tolerance in transgenic plants. Heat stress can cause proteins, RNA and other molecules to fold incorrectly, affecting their assembly, translocation, turnover and activity. Heat-shock proteins (HSPs) and other so-called 'chaperones' that keep proteins and RNA in their correct conformation are expressed to mitigate this problem and these have therefore also attracted much attention. The expression of HSPs is under the control of HSFs.

The key question for plant biotechnology is will the manipulation of any of these genes provide a consistent improvement in drought and/or heat tolerance under field as opposed to laboratory conditions. At present there are no transgenic crop varieties being marketed on the basis of improved drought or heat tolerance, but all of the major plant biotechnology companies claim to have such varieties in development. Monsanto appears to be leading the way with plans to market drought-tolerant maize possibly as early as 2012. The exact nature of the traits involved is not yet in the public domain and there may be more than one; Monsanto has experimented with over-expression of transcription factors and RNA chaperones. The yield increases that are claimed are relatively modest, from 6.7–13.4% under drought conditions. Nevertheless, these varieties would hopefully represent the start of a process of long-term improvement.

3.12 Salt Tolerance

Salt stress is linked with drought stress in both a plant physiological and a practical sense. In a plant physiological sense, both drought and salt impose an osmotic stress on the plant. In a practical sense, farmers respond to long-term drought by irrigating, usually with river water, and while river water is 'fresh' it does contain low concentrations of salt, with salt concentration rising nearer the sea. As the water evaporates from the land it leaves the salt behind and the salt concentration in the soil gradually increases until it begins to affect plant growth and crop yield. This problem was recognised even in ancient times; the ancient Egyptians, for example, were aware that continuous irrigation of farmland with water from the Nile would affect soil fertility. It is now estimated that more than a third of all of the irrigated land in the world is affected by salinity.

High salt concentration in the soil prevents plants from taking up water. It also causes a build-up of Na^+ and Cl^- ions and skews the K^+/Na^+ ratio. Plants respond by restricting the uptake of salt, adjusting their osmotic pressure by synthesising compatible solutes such as proline, glycinebetaine and sugars, and sequestering salt in the cell vacuoles. They also neutralise reactive oxygen species that are generated during the stress response.

Clearly, the development and use of crops that can tolerate high levels of salt in the soil would be part of a practical solution to the problem. As with drought and heat stress, however, progress has been hampered by the complexity of the underlying genetics. Consequently, only a few cultivars of the major crop species are registered as salt-tolerant. However, the existence of salt-tolerant plants (called halophytes) and differences in salt tolerance between genotypes within salt-sensitive plant species (called glycophytes) indicate that there is a genetic basis to salt tolerance that could be exploited if it were understood.

As with drought and heat stress, plants respond to salt stress through the action of multiple signalling pathways involving protein kinases and transcription factors. Some of these pathways overlap or interact with the pathways induced by drought stress. There has been some success in improving salt tolerance, at least in the laboratory, through the manipulation of the genes involved in these pathways. However, more progress has been made through the introduction of novel genes or the manipulation of expression of native genes to impart ion homeostasis, osmotic regulation or antioxidant protection.

Na^+ extrusion from plant cells is powered by an H^+-ATPase which generates an electrochemical H^+ gradient that allows Na^+/H^+ antiporters (effectively molecular pumps) to pump Na^+ out of the cell. The over-expression of one of these antiporters, SOS1, in transgenic Arabidopsis has been shown to improve salt tolerance under laboratory conditions. There are also Na^+/H^+ antiporters that pump Na^+ out of the cytosol into the vacuole, where they do not cause damage. The over-expression of one of these, NHX1, in transgenic Arabidopsis and tomato has also been shown to result in plants that are able to grow in high salt concentrations. This work was pioneered by Eduardo Blumwald at the University of California.

The tomato plants over-expressing NHX1 could grow, flower and set fruit in the presence of 200 mM NaCl. Importantly, while the leaves accumulated high sodium concentrations, the fruits did not and were edible. The technology was subsequently applied to oilseed rape, with similar results. Leaves of the GM plants grown in the presence of 200 mM NaCl accumulated sodium to up to 6% of their dry weight, but the seed yields and oil quality were not affected. Other researchers have since had similar success with rice and maize.

A complementary strategy for improving salt tolerance concerns the synthesis and accumulation of a class of osmoprotective compounds known as compatible solutes. These are relatively small, organic metabolites, including some amino acids, such as proline, and their derivatives, such as glycinebetaine (a small, trimethylated amino acid), polyols (alcohols with multiple hydroxyl groups) and sugars. The accumulation of these osmotically-active compounds in the cytosol increases the osmotic potential to balance the high concentration of Na^+ and Cl^- ions outside the cell. GM approaches have focused on increasing proline and glycinebetaine synthesis. Increased proline accumulation has been induced by over-expression of genes encoding pyrroline-5-carboxylate (P5C) synthase (P5CS) and P5C reductase (P5CR), the two enzymes that catalyse the synthesis of proline from glutamic acid, and by reducing the expression of proline dehydrogenase, which is involved in proline breakdown. However, success has been limited by lack of understanding of the regulation of proline synthesis, metabolic limitations on levels to which proline can accumulate and a weak correlation between the concentration of proline and the salt tolerance that has been achieved.

Increasing glycinebetaine accumulation has also proved difficult, hampered by problems of metabolic fluxes and cellular compartmentation of the substrate and product. Plants that naturally accumulate glycinebetaine include spinach and sugarbeet; in these species, synthesis of glycinebetaine occurs in the chloroplast, catalysed by choline mono-oxygenase (CMO) and betaine aldehyde dehydrogenase (BADH). Bacteria have two different pathways, one involving choline dehydrogenase (CDH) and BADH, the other a single enzyme, choline oxidase (COD, or COX). Engineering the second of these into GM plants is attractive because it involves only a single gene and this has been done in Arabidopsis, *Brassica* and tobacco (as a model) using genes from *Arthrobacter globiformis* or *Arthrobacter panescens*. In some cases the gene was first modified so that the encoded protein would be transported to the chloroplast but improved salt tolerance was achieved in the GM plants with or without this modification. However, the degree of tolerance that was achieved was variable and glycinebetaine could not be induced to accumulate to levels comparable with those in plants that accumulate it naturally. As with proline, a better understanding of the metabolic constraints on glycinebetaine accumulation is required if more progress is to be made.

To conclude this section, it is clear that improved salt tolerance is possible; indeed, it has already been achieved. However, Blumwald reported his successful work on tomato in 1999 and 11 years later there is still no commercial use of the technology. Part of the problem may be that the market is seen to be limited: most crops are not grown on salt-affected soils. Biotech companies may therefore be concerned that they will not get a return on the investment required for developing salt-tolerant varieties. If that is the case, we are unlikely to see such varieties on the market while the cost of development and regulatory compliance is so high.

3.13 Biopharming

Biopharming is the term applied to the use of GM plants to produce pharmaceuticals, vaccines or antibodies. It has also been applied to the production of proteins for industrial uses. Many vaccines and other pharmaceutical products are already produced in GM microbes, of course. Indeed, insulin was first produced in GM bacteria as long ago as 1981, enabling people with diabetes to be treated with human insulin instead of insulin from pigs or cattle. In 2007, a Canadian company, Sembiosys, announced that it had genetically modified safflower to produce insulin in its seeds. The attraction of using plants is that growing plants in the field would be relatively cheap compared with growing bacteria in sealed vats, and potentially would allow production on a much larger scale. Sembiosys points out that demand for insulin is increasing rapidly with the rise in cases of diabetes and the development of new delivery methods such as inhalation that require more insulin. However, if it is to be successful this product will have to compete with a well-established and widely accepted production system.

The possibility of producing vaccines in GM plants has also caused great excitement, although progress has been slow since the idea was first put forward in the 1990s. It is estimated that there are 12 billion injections administered worldwide every year. Thirty per cent of these are not performed under sterile conditions. Furthermore, millions of people die every year from diseases that are preventable through vaccination. People may not have access to vaccines for one or more of several reasons. They may live in remote places that lack the infrastructure (transport, refrigeration, availability of needles and syringes) to deliver and store the vaccine. If they have to walk long distances to a clinic to have their children and themselves vaccinated they may find it impossible to repeat the journey for a subsequent booster vaccination, so they are not protected adequately. There may not be sufficient numbers of medical professionals to administer a vaccination programme or the vaccine may just be too expensive. The production of vaccines in GM plants could potentially address some of these problems.

One of the first reports of successful vaccine production in a GM plant was for hepatitis B. Hepatitis B causes acute and chronic disease of the liver and is associated with liver failure and liver cancer. It is a big killer throughout the developing world. The first hepatitis B vaccine was a protein (the surface antigen) extracted from the blood of people infected with the disease and was produced in the 1970s. As HIV spread in the 1980s, this practice was viewed as too dangerous and the gene that encoded the protein was engineered into GM yeast to produce the vaccine. The vaccine has been made this way ever since and is extremely effective.

Unfortunately it is too expensive for many developing countries to afford and even in the United Kingdom the vaccine is only offered to those people who are considered to be at special risk of contracting the disease.

A group of researchers led by Hugh Mason at Cornell University genetically modified potato plants to make the hepatitis B surface antigen in their tubers. The vaccine can be administered orally (patients eat a piece of the GM potato) and has been shown to work in animal and human trials. The original intention was that people would be able to grow plants producing the vaccine themselves. However, this would make the control of dosage impossible and is not now considered to be a realistic option. Nevertheless, the advantages are clear: the potato can be stored at room temperature for a long time, and even moderate-scale production by agricultural standards could supply enough vaccine to meet global demand. Similar experiments have been performed to produce a vaccine for pathogenic *Escherichia coli*, a bacterium that causes severe, sometimes fatal food poisoning, in potato and tobacco. However, the first human trials of this vaccine were completed in 1997 and there has been little sign of progress since.

These vaccines rely on a pathogen-derived protein or part of a protein to stimulate immunity. These so-called subunit vaccines cannot cause disease but may not always elicit the entire range of responses necessary to provide effective immunity. The hepatitis B vaccine, for example, used the HbsAg surface antigen, while the pathogenic *Escherichia coli* vaccine used the heat-labile toxin B subunit (LT-B). Other vaccines of this sort produced in plants include a rabies virus antigen in tomato, a cholera antigen in tobacco and potato, the Norwalk virus capsid protein in tobacco and potato, and a human cytomegalovirus antigen in tobacco.

The technology has also been applied to viruses that affect livestock. For example, the company ProdiGene, based in Texas, has produced an edible vaccine for transmissible gastroenteritis virus (TGEV) in pigs, while the foot and mouth disease virus (FMDV) antigen, VP1, has been expressed in Arabidopsis, potatoes and alfalfa.

A different approach to the production of vaccines in plants is the use of vectors based on plant viruses and this has been used in particular for the production of vaccines for veterinary use. This approach has the advantages that very high levels of vaccine production can be achieved, viral genomes are small and relatively easy to manipulate, and infecting plants with a modified virus is easier than genetic modification of the plant. On the other hand, the production of the vaccine is transient, not heritable, there are limitations on the size and complexity of the DNA sequences that can be expressed in this way, and the genetic modification of viruses, rightly or wrongly, raises safety concerns.

One use of modified viruses has been to insert a DNA sequence encoding an antigenic peptide into the viral coat protein gene so that the peptide is expressed on the surface of assembled virus particles, often called chimaeric virus particles (CVPs). Antigenic proteins in multiple copies on the surface of a CVP are often much more effective at eliciting an immune response than when presented on their own. The modified viral particles are purified from the plant tissue for administration to animals. Alternatively, a gene is introduced into the viral genome so that it is expressed in infected cells, usually to make the protein on its own rather than fused with anything else. The protein may then be purified for administration, or animals could be fed plant material directly.

Plant viruses used in this way include *Cowpea mosaic virus* (CPMV), *Tobacco mosaic virus* (TMV), *Tomato bushy stunt virus* (TBSV), *Alfalfa mosaic virus* (AlMV), *Plum potyvirus* (PPV) and *Potato virus X* (PVX). Examples of promising results with the technique include protection of mink and dogs from canine parvovirus (CPV) by injection of a CPMV chimaera containing a short peptide from a CPV coat protein (1999). Mice have also been immunised against *Mouse hepatitis virus* (MHV) by injection with chimaeric TMV particles containing a peptide from an MHV coat protein. In another experiment, mice were immunised against rabies with an AlMV chimaera.

The potential of vaccines produced in this way seems huge. However, there are problems to be overcome. These include scaling-up production, addressing safety concerns, negotiating regulatory hurdles, demonstrating that vaccines produced in plants are as effective as those produced by other means, and engaging pharmaceutical companies in plant vaccine development. Vaccines produced in plants may be administered by injection but it was the notion of edible vaccines that caused much of the early excitement around plant-derived vaccines. It is now recognised that the notion of edible vaccines may be difficult to make a reality in practice, because of problems with controlling dose and the requirement for an adjuvant (a protein that stimulates the immune response and increases the effectiveness of a vaccine). Possible adjuvants are themselves derived from pathogenic bacteria and the most effective are toxic. There has been some success in recent years in using mutagenesis to produce non-toxic forms of some adjuvants and expressing these in plants, but the technology still requires some development and safety testing. Vaccines administered orally or nasally also present the difficulty of oral tolerance, which is a condition of unresponsiveness that can arise after a vaccine is administered in this way.

Plants have also been used to make antibodies. Antibodies bind target proteins with an extremely high degree of specificity and have a variety of uses in medicine, including disease diagnosis and the treatment of some cancers and autoimmune conditions. Their high degree of specificity means that they are also being developed for a number of uses outside medicine.

The production of antibodies in plants is not a trivial matter, partly because antibodies are complex proteins comprising multiple subunits. Full-length serum antibodies comprise two identical heavy (H) and two identical light (L) chains, and specificity is imparted by variable regions at one end of each chain. Secretory antibodies are even more complex, comprising dimers of the serum antibody structure plus a joining (J) chain and a secretory component (SC). This can be got around to some extent because less complex derivatives retain specificity. There are a number of different types of these derivatives, one example being single-chain Fv molecules (scFvs), which comprise the variable regions of the heavy and light chains of an antibody, held together by a flexible linker. However, antibodies are also glycosylated, in other words they have oligosaccharides attached to them, and the mechanisms of glycosylation in plants are different to those in animals. As a result, antibodies produced in normal plants are immunogenic in mammals. There has been progress in producing GM plants that glycosylate proteins in the mammalian manner, notably at the Universität für Bodenkultur in Vienna. These plants have been referred to as 'humanised' with respect to their glycosylation mechanisms and have been used to produce monoclonal antibodies that are equivalent to antibodies produced in mammalian cells.

One of the first and most impressive uses of plants for antibody production was the production in tobacco of a monoclonal antibody, Guys 13, which binds to the surface protein of *Streptococcus mutans*, the bacterium that causes tooth decay. Application of this monoclonal antibody to teeth prevents colonisation by *S. mutans*. Production of the antibody in plants was achieved by Julian Ma at Guys Hospital in London. The four subunits, H, L, J and SC, were each expressed in separate transgenic tobacco plants which were then crossed to produce a hybrid containing all four transgenes. The Guy's 13, so-called 'plantibody' technology, is licensed to Planet Biotechnology Inc. and is undergoing clinical trials under the product name CaroRx™.

Another project that looks promising is the expression in GM soybean of a humanised antibody that recognises the herpes simplex virus (HSV)-2 glycoprotein B. The plant-expressed antibody has been shown to protect mice against vaginal transmission of HSV-2 and its development is continuing. An example of the potential use of an antibody derivative is cFv 84.66, which recognises carcinoembryonic antigen (CEA), a protein that is present at elevated levels in people with colorectal, gastric, pancreatic, lung and breast cancer. The antibody could potentially be used to facilitate early diagnosis of these cancers, or to direct therapeutic drugs to cancer cells in the body (for example it has been fused with interleukin-2 for this purpose). It has been produced in rice, wheat, tomato, pea and tobacco plants.

Clearly the production of vaccines and antibodies is difficult and complex. However, the term biopharming has also been applied to less challenging applications of GM, such as the production of enzymes for use in industry. One example of this is the production of trypsin, an animal protease (an enzyme that breaks down proteins), that has a variety of applications in biological research and is also used in the food industry, for example to predigest baby food. The company ProdiGene has genetically modified maize to produce trypsin. The product, TrypZean, is already on the market for use in laboratory applications.

3.14 Removal of Allergens

Food allergy is an increasingly serious medical problem. The Food Standards Agency of the United Kingdom reported in 2000 that around 1.4–1.8% of the UK population and up to 8% of children in the UK suffer from some type of food allergy. It appears that there have been rapid increases in the number of incriminated foods and the frequency of severe reactions. At present there is no specific treatment for food allergy, apart from dietary avoidance, although patients at risk of anaphylactic shock carry adrenaline in case of accidental exposure. Avoidance of allergens in processed foods can be difficult because their presence may not be obvious and products may not be labelled properly. The risk of accidental exposure, for example in a restaurant meal, is a constant anxiety that affects quality of life significantly.

Ironically, the possibility that some genes used in plant biotechnology could encode allergenic proteins has concerned consumers and regulatory authorities. This is considered in Chapters 4 and 5. The other side of the coin is that genetic modification could be used to remove allergens from the food chain.

Well known plant allergens include the 2S albumins, a group of storage proteins found in the seeds of legumes (peas and beans), crucifers (cabbages, turnips, swedes, mustards and radish) and many nuts. As well as acting as food stores for the seed, some inhibit digestive enzymes, perhaps protecting the seeds from being eaten, and some have antifungal properties.

Another group of proteins that includes several allergenic members are lipid transfer proteins, a family of small proteins present in seeds and other parts of a diverse range of plants, including fruits, oilseeds and cereals. They cause sunflower allergenicity in southern European populations, resulting in severe symptoms, including anaphylaxis, albeit in a small number of sufferers. The function of lipid transfer proteins is not known but they have been associated with plant defence.

Another family of proteins involved in plant defence are the PR (pathogen-related) proteins. These are made by plants in response to microbial infection. They include a number of allergenic proteins and are responsible for the allergenicity of chestnut, avocado, birch pollen, apples, cherries, celery and carrots.

Baker's asthma, a respiratory allergy caused by inhalation of wheat flour, is common in workers in the flour milling and baking industries. The allergenic proteins responsible appear to be small inhibitors of two digestive enzymes, α-amylase and trypsin. They are present in rye, barley, maize and rice as well as wheat.

The obvious target for biotechnologists is to remove these proteins from crop plants using gene suppression techniques (Chapter 2). These methods can lead to almost total suppression of gene activity and, while it might not be sufficient to make a food completely safe for someone who is allergenic to it, it could reduce the frequency and severity of reactions caused by accidental exposure. However, clearly this can only be done if the protein does not have an essential function in the plant.

There are currently only a few examples of experiments in which this has been tried. In one, the amounts of α-amylase inhibitors in rice were reduced substantially. There are at least ten different genes encoding these proteins in rice and to affect them all using conventional plant breeding or mutagenesis would be impossible. In another experiment, antisense technology was used to reduce the levels of the Lol p5 allergen in the pollen of ryegrass. The pollen of the GM ryegrass was shown to have reduced allergenicity.

3.15 Conclusions

Genetic modification is now an established technique in plant breeding, complementing traditional methods and other new developments such as genomics-based accelerated breeding. This chapter has described successful applications of GM in agriculture, some failures and some exciting prospects for the future. It is clear that GM is already bringing significant benefits to agriculture wherever GM crops are allowed to be grown. However, GM will only realise its full potential if the legislation covering the use of the technology does not make it too difficult or expensive to produce and market GM crops and consumers accept GM crop products. These issues are discussed in the next two chapters.

4 LEGISLATION COVERING GM CROPS AND FOODS

4.1 Safety of GM Plants Grown in Containment

The question of GM crop and food safety was first considered in the United Kingdom in the early 1980s when the first GM plants were being produced. Responsibility for ensuring that the technology was developed safely was given to the Health and Safety Executive (HSE), a government agency responsible for the regulation of almost all of the risks to health and safety associated with the workplace in the United Kingdom. The HSE had already set up a committee, the Advisory Committee on Genetic Modification (ACGM), to control the use of GM micro-organisms. The responsibility of this committee was extended to cover the production and use of all GM organisms (GMOs) in containment. The ACGM is now known as the Scientific Advisory Committee on Genetic Modification (Contained Use) (SACGM (CU)), and it provides technical and scientific advice on 'all aspects of the human and environmental risks of the contained use of genetically modified organisms'.

No organisation in the United Kingdom can produce or hold GMOs without the permission of the HSE. The HSE ensures that any organisation that proposes working on GMOs has the facilities required and has staff trained and experienced in the handling and disposal of the organisms and contaminated waste. Different levels of containment are required for different GMOs, depending on the risk that they represent.

An organism is said to be in containment if physical, chemical or biological barriers are used to limit contact between it and other organisms or the environment. With bacteria, biological containment is often achieved by using strains that are genetically disabled and therefore unable to survive outside the laboratory. It can be applied to plants by using species that do not survive in the local environment and do not cross with native wild plant species (wheat, for example, does not survive outside agriculture and has no wild relatives in the UK), or by adopting measures such as the removal of flowers before pollen is shed. An example of a chemical barrier to the spread of a GM plant is the use of a herbicide, although it would be more likely to be applied in the post-harvest treatment of a field test site than in strict containment conditions. Physical containment involves the use of specially designed laboratories, plant growth rooms and glasshouses. There are four categories of containment for GM micro-organisms and two for GM plants and animals (Table 4.1). In the case of GM plants the legislation only covers their use in the laboratory and glasshouse; field releases are covered separately.

Even for GMOs that are considered to be of no risk to human health or the environment, the laboratory in which they are held and used must satisfy basic standards. For example, the laboratory must be easy to clean, bench-tops and floor must be sealed and if the laboratory is mechanically ventilated the air flow must be inwards. Hand-washing facilities must be provided next to the exit and basic protective clothing, such as a lab-coat and disposable gloves, must be worn and removed before leaving the laboratory. Access to the laboratory must be restricted. Any procedures that produce aerosols must be done in a safety cabinet, effective disinfectants must be available next to every sink, spillages must be dealt with immediately and recorded, bench tops must be cleaned after use and a good general standard of cleanliness must be maintained. All contaminated glassware must be stored safely and sterilised after use, and all contaminated waste must be stored safely in a designated bin and autoclaved (sterilised by heat and pressure) before disposal.

Table 4.1 Risk categories of GM micro-organisms, plants and animals kept in containment in the UK.

	GM micro-organisms				GM plants and animals	
	Class 1	**Class 2**	**Class 3**	**Class 4**	**A**	**B**
Risk to human health or environment	None or negligible	Low	Medium	High	Less risky than non-GM parent	More risky than non-GM parent
Notification requirement	HSE must be notified that GM work is to be undertaken	HSE must be notified of each activity in these categories			HSE must be notified that GM work is to be undertaken	HSE must be notified of each activity
Containment and safety measures	→ Increasing requirement for measures to contain organism and protect handler				→ Increasing requirement for containment measures	

At the heart of the safe handling of GMOs is expert risk assessment. All projects involving GMOs have to be risk-assessed by the project leader and the risk assessment must be considered by an organisation's internal Genetic Modification Safety Committee. The project leader is required to give information on the experience level of staff who will work on the project and the training that they have received. For a project concerning GM plants, information has to be given on the host plant and the gene or genes being inserted. To assess any risk to human health the project leader has to consider any possible induction of or increase in toxicity and/or allergenicity compared with the parent plant and the risk of accumulation of toxicity through food chains.

The project leader also has to assess any risk posed by the proposed GM plants to the environment, particularly the potential of the plants to be more 'weedy' than the parent plant. This assessment includes factors such as colonisation ability, seed dispersal mechanisms, resistance to control measures such as herbicides, increased toxicity to insects and other grazers and any other possible change in the plants' interaction with their environment. The potential for and consequences of the sexual transfer of nucleic acids between the GM plants and other plants of the same species or a compatible species has to be considered, particularly if the plants have the ability to transfer novel genetic material to UK plant species. The risk and consequences of horizontal gene transfer to unrelated species, for example by a virus, bacterium or other vector, also has to be taken into account. Finally, an assessment has to be made of the potential of the GM plants to cause harm to animals or beneficial micro-organisms.

This risk assessment has to be completed before the experiment begins and despite the fact that the plants will be kept in containment in specially designed greenhouses or controlled environment rooms. These have filtered negative air pressure ventilation, sealed drains and a chlorination treatment system for drainage water to ensure that no viable plant material escapes into the environment.

The proper risk assessment of experiments involving GMOs and their safe handling and containment are legal requirements and failure to observe the regulations could result in prosecution. It would certainly result in serious cases in the loss of an institution's licence to work with GMOs. All institutions that work with GMOs are inspected regularly by the HSE.

4.2 Safety of Field Releases of GM Plants

In the late 1980s it seemed that the growing of GM plants in the field in the United Kingdom was likely to become widespread and the government decided that it would have to be controlled. Regulations covering the field release of GM plants and animals were written into the Environmental Protection Act of 1990 and a statutory advisory committee was set up to provide advice to government regarding the release and marketing of genetically modified organisms. This committee was called the Advisory Committee for Releases into the Environment (ACRE). The regulations were updated through the GMO Deliberate Release Regulations 2002 which implemented the European Union's directive on the deliberate release of GMOs, Directive 2001/18/EC.

ACRE advises the Secretary of State for Environment, Food and Rural Affairs, Scottish Ministers, the Welsh Assembly Secretaries and the Department of the Environment in Northern Ireland. It also advises the Health and Safety Executive on human health aspects of releasing GMOs. ACRE is made up predominantly of academics, and current members have expertise in agronomy, biodiversity, ecology, entomology, microbiology, plant molecular biology, plant breeding, plant physiology and farming practice.

Anyone who wants to undertake a field release of a GM plant in the UK has to obtain permission from the government Department for Environment, Food and Rural Affairs (DEFRA). A detailed risk assessment of the plants must be undertaken and considered by ACRE, who then advise DEFRA on whether or not to allow the release to go ahead. More information will be available for this risk assessment than the one described for a proposal to make a set of GM plants and grow them in containment because the plants will already have been made and studied. A typical risk assessment will include:

- Information on the host plant species:
    - The full name of the plant species and the breeding line used.
    - Details of the sexual reproduction of the plant, generation time and the sexual compatibility of the plant with other cultivated or wild plant species.
    - Information concerning the survivability and dissemination of the plant and the geographical distribution of the species.
- A description of the methods used for the genetic modification and the nature and source of the vector used to modify the plant.
- Information on the nature of the genetic modification:
    - Details of the novel genes introduced into the plant, including size, intended function and the organisms from which they originated.
    - Information on when and where in the plant the novel gene or genes is/are active and the methods used for finding out.
    - Information on the location of the inserted novel DNA in the plant cells and the number of copies of the novel gene or genes that are present.
    - The size and function of any region of the host plant genome that has been deleted as a result of the genetic modification.
    - An analysis of the genetic stability of the novel gene or genes.
- An assessment of the GM plants:
    - A description of the trait or traits and characteristics of the genetically modified plant which have been introduced or changed.
    - An assessment of any differences between the genetically modified plant and its parent in respect of methods and rates of reproduction, dissemination and survivability.
    - A description of detection and identification techniques for the genetically modified plants.

- An assessment of potential risks posed to health and/or the environment:
  - An assessment of any potential toxic effects on humans, animals and other organisms.
  - An assessment of the likelihood of the genetically modified plant becoming more persistent than the recipient or parental plants in agricultural habitats or more invasive in natural habitats.
  - A description of the mechanism of interaction between the genetically modified plants and target organisms (for example if the plants have been engineered to be resistant to insects) and any potentially significant interactions with non-target organisms.
  - An assessment of the potential environmental impact of the interaction between the genetically modified plant and target or non-target organisms.
  - An assessment of the potential for transfer of genetic material from the genetically modified plants to other organisms.
  - Any selective advantage or disadvantage conferred to other species that may result from genetic transfer from the genetically modified plant.
  - Information about previous releases of the genetically modified plants.
- Information on the release site:
  - The location and size of the release site or sites.
  - A description of the release site ecosystem, including climate, flora and fauna.
  - Details of any sexually compatible wild relatives or cultivated plant species present at the release sites.
  - The proximity of the release sites to officially recognised protected areas that may be affected.
- A description of the management of the field trial:
  - The purpose of the trial.
  - The foreseen dates and duration of the trial.

  - The method by which the genetically modified plants will be released.
  - The method for preparing and managing the release site, including cultivation practices and harvesting methods.
  - The approximate number of genetically modified plants (or plants per m^2) to be released.
- A description of containment measures to be adopted during and after the field trial:
  - A description of any precautions to maintain the genetically modified plant at a distance from sexually compatible plant species and to minimise or prevent pollen or seed dispersal.
  - A description of the methods for post-release treatment of the site. These are likely to include ploughing of the site, irrigation to encourage germination of any seed in the soil, the removal of any plants that sprout by spraying with an appropriate total herbicide and a period of one to two years when the site is kept fallow and monitored.
  - A description of how the genetically modified plant material will be disposed of.
  - Details of how the site is to be monitored after the trial is over.
  - A description of emergency plans in the event that an undesirable effect of the modification becomes evident during the trial or that the plants spread.
  - An assessment of the likelihood and consequences of theft of GM material from the trial, vandalism of the trial, or accidental movement of GM material off the trial site by any means (e.g. on field machinery).

Clearly, providing all of this information is a lengthy and expensive business. On top of that, the cost of a new application is £5000 with each actual release costing an additional £850 per year; not insignificant in the context of research budgets. Applicants have to wait at least 90 days for a decision, with the clock stopped if the applicant is asked for further details until the information is provided. This delay can be a problem because, obviously, seed has to be sown at a particular time of year. Applicants are also required to provide the map reference of the trial site and place this in the public domain through an advert in the local press notifying the public of the trial. As a result, many field trials in the UK have been vandalised and most companies and institutions have decided that it is not worth attempting to run them: there were no GM crop field trials at all in the UK between 2004 and 2009.

In contrast, tens of thousands of field trials of GM plants have been performed in the USA on species from soybean to walnut. The vast majority of these have caused no problems and the American authorities have made the application procedure much easier as a result. However, they are discovering that with some GM plant experiments, such as those involving plants that make pharmaceutical products, they have to be more careful (see Chapter 5). Nevertheless, the fact that field experiments are so much more difficult in the UK and Europe than in the USA does put European researchers at a competitive disadvantage. Despite the fact that no problems have arisen from the field experiments of GM plants that have taken place in Europe, there is no indication that the regulations will be relaxed in the foreseeable future.

4.3 Safety of GM Foods

The UK government can take advice on the safety of foods derived from GM crops from another committee, the Advisory Committee on Novel Foods and Processes (ACNFP). ACNFP was established in 1988 to advise the responsible authorities in the United Kingdom on any matters relating to novel foods and novel food processes. Currently it comprises 15 members, most of whom are academics, with expertise in allergenicity, genetics, immunology, microbiology, nutrition and toxicology. It comes under the jurisdiction of the Food Standards Agency (FSA).

The assessment made by the ACNFP essentially follows guidelines endorsed by the World Health Organisation (WHO) and the same guidelines are followed by almost all regulatory authorities around the world. At the heart of the process is the concept of substantial equivalence. This concept is often attacked by anti-GM campaigners who claim that it merely means showing that a GM plant or food derived from it is roughly the same as its non-GM counterpart. In reality substantial equivalence involves a comprehensive biochemical and molecular comparison of a GM food and its conventional equivalent and a detailed analysis of any differences.

The fact is that very few foods consumed today have been subject to any toxicological studies, yet they are generally accepted as being safe to eat. The difficulties of applying traditional toxicological testing and risk assessment procedures to whole foods, GM or otherwise, makes it pretty well impossible to establish absolute safety. The aim of the substantial equivalence approach, therefore, is to consider whether the genetically modified food is as safe as its traditional counterpart, where such a counterpart exists.

The process begins with a comparison between the GM plant or food and its closest traditional counterpart in order to identify any intended and unintended differences. These differences then become the focus of the safety assessment and, if necessary, further investigation. Factors taken into account in the safety assessment include:

- The identity and source of novel genes (in particular is the source a well-characterised food source or is it entirely new to the food chain).
- Composition of the plant and/or food derived from it compared with its traditional counterpart.
- Effects of processing/cooking.
- The methods used to make the GM plant.
- The stability and potential for transfer of the novel gene or genes.
- The nature of the protein encoded by the novel gene or genes.
- Potential changes in function of novel genes and proteins.
- Potential toxicity of novel proteins.

- Potential allergenicity of novel proteins.
- Possible secondary effects resulting from expression of the novel gene or genes, for example by disruption of a gene in the host plant, knock-on effects on metabolic pathways and changes in the production of nutrients, anti-nutrients, toxins, allergens and physiologically active substances.
- Potential intake and dietary impact of the introduction of the genetically modified food.

The technology available to undertake these sorts of studies has moved on tremendously in the last few years. The activity of thousands, sometimes tens of thousands of genes can be determined in a single experiment. Similarly, the amounts of thousands of proteins and metabolites present in a plant can be measured. These techniques of transcriptomics, proteomics and metabolomics are revolutionising safety testing. However, it is possible, likely even, that changes in gene expression, protein synthesis and metabolite profiles would be found in new varieties of crops produced by any of the methods used in plant breeding, whether it be crossing, mutagenesis, genetic modification or anything else. The difficulty will be in identifying those changes that are significant and making rational decisions about them.

Some work has already been done to compare the effects of GM and other factors on wheat seed composition at the metabolomic centre at Rothamsted Research, UK (MeTRO). The substantial equivalence of three GM wheat lines and their non-GM parents were examined using metabolite profiling of samples grown in field trials on sites near Bristol, in the west of England, and at Rothamsted itself in the east, over three years. The conclusions of the study were that the differences in grain composition caused by site and year of cultivation were greater than those caused by genetic modification. A similar study of global gene expression (transcriptomics) in transgenic and conventionally bred wheat lines showed that the differences in gene expression between conventionally bred genotypes were much larger than those between GM and non-GM genotypes.

4.4 European Union Regulations

The European Union's directive, GM Food and Feed Regulation (EC) No. 1829/2003, was adopted in 2004, bringing the regulation of GM crop use and release under the control of the European Commission. The EU recognises two different types of field release of GM crops, one for research purposes only (a Part B release) and one for commercial release (a Part C release). Consent for a Part C release may be granted for cultivation, food and feed use, or for food and feed use alone (i.e. for import and consumption in the European Union but not for cultivation). Permission for a Part B release can be granted by an individual Member State, as described above for the UK. However, applications for a Part C release anywhere in the European Union have to be approved by the European Commission.

Each new application is assessed by the European Food Safety Authority (EFSA). If EFSA approves the application, it is voted on by a working group with representatives from all 25 Member States. The UK representative may take advice from ACRE, ACNFP and another UK advisory committee, the Advisory Committee for Animal Feedingstuffs (ACAF) as well as considering the assessment of EFSA. The European Commission has adopted a complicated system of Qualified Majority Voting (QMV), meaning that it is relatively easy for a minority of countries to block an application. If that happens, the Council of Europe and even the Council of Ministers may become involved in a series of futile cycles.

BASF's Amflora potato variety (Chapter 3) was bogged down in this process for ten years, which is a farcical situation. Indeed, the European Commission failed to approve a single application between 1998 and 2004 because six Member States (France, Italy, Denmark, Greece, Austria and Luxembourg) managed to block every one. This *de facto* moratorium undoubtedly damaged trade relations and stymied the development of the plant biotechnology industry in Europe. The impasse was finally broken in May 2004 when the European Commission authorised

Syngenta Bt-11 sweetcorn, an insect-resistant and gluphosinate-tolerant variety, for import and food use. Later in 2004 it extended the consent for Mon810, an insect-resistant variety that had been authorised for food use throughout the EU and for cultivation in France and Spain since 1998. The extension allowed for cultivation of this variety, which has been quite successful in Spain, throughout the EU. Note that individual Member States still impose restrictions, however. France, for example, banned the cultivation of this variety in 2008 after it had appeared to be becoming popular with farmers in 2006 and 2007 (Figure 3.3).

The log-jam in applications started to shift but progress was still slow and Austria, France, Germany, Luxembourg and Greece retained their own national bans in defiance of the European Commission. Frustrated with the lack of progress, the USA, Australia, Argentina, Brazil, Canada, India, Mexico and, surprisingly, New Zealand, complained to the World Trade Organisation (WTO) and in 2006 the WTO ruled that the European Union's position was illegal and also criticised the bans imposed by individual Member States. Austria and Hungary are now the only Member States that still ban GM crops and their products outright.

It remains extremely difficult to obtain Part C consent for cultivation and marketing of GM crops and their products in the European Union and the area of GM crops in Europe is still in the hundreds of thousands of hectares (Figure 3.3) in contrast to the tens of millions of hectares in the Americas and Asia. It is easier, although still problematic, to gain approval for food and feed use alone without seeking approval for cultivation. As a result, efforts to develop new varieties for cultivation in Europe have all but been abandoned by biotech companies. Instead, companies are focusing on obtaining permission for food and feed use so that their potential customers elsewhere in the world can be reassured that the European market is open to their products. This means that European farmers are competing with GM crops but are unable to use them.

The European Commission has recently proposed that some of the decision-making on GM crop and food issues should be devolved back down to national Member State governments, but details on how or when this is going to happen are not available at the time of writing.

4.5 Labelling and Traceability Regulations

The first GM plant product to come onto the market in the United Kingdom was paste made from slow-ripening tomatoes. This product was approved for food use in the UK by ACNFP in 1995, but the product has not been available since 1999. The paste was sold through two major food retailers, Sainsbury and Safeway, clearly marked with a large label stating that it was made from GM tomatoes. The UK food retail industry intended to pursue this policy for all foods derived from GM plants, at least until consumers were familiar with the new technology. However, these plans were thrown into confusion in late 1996 when shipments of that year's harvest of soybean and maize imported from the USA arrived. Both contained at that time approximately 2% GM material with the GM and non-GM all mixed together. With no legislation covering the labelling of GM foods, no agreement with the Americans to supply segregated GM and non-GM produce and insufficient time to identify alternative suppliers (the industry appeared to have been taken by surprise), retailers had a choice: either label everything containing US soybean and maize as potentially containing GM material or abandon the labelling policy. It chose the latter, a decision that, with hindsight, was undoubtedly a mistake because consumers felt that GM food was being introduced behind their backs.

Labelling controls covering GM foods were finally introduced in Europe in 1997 but not ratified in the UK until 1999. The regulations required that any food containing material from GM crops had to be labelled, with the following exemptions:

- Refined vegetable oils, sugar and other products that do not contain DNA or protein.

- Foods that contain small amounts (below 1%) of GM material as a result of accidental mixing.
- Food sold in restaurants and other catering outlets (the UK opted out of this exemption and required caterers to provide written or verbal information covering GM foods to their customers).
- Animal feed.

The regulations were updated through directives on the regulation of GM food and feed (1829/2003) and the traceability and labelling of GMOs (1830/2003). These regulations, which came into force in April 2004, require the labelling of all food and animal feed containing GMOs or GMO-derived ingredients; so animal feed is no longer exempt. The threshold for labelling is reduced to 0.9% for the accidental presence of GM material, with zero tolerance if the GM material has not been approved for use in Europe. This zero tolerance rule for products that have been approved elsewhere but are not yet approved in Europe is becoming increasingly unworkable as the amount of GM material in imported food and feed increases. The traceability and labelling regulations require a system for tracking GM products through the supply chain from seed company to farm, processors, distributors and retailers. The exemptions for refined products no longer apply, but the regulations do not cover products from GM-fed animals or enzymes produced in GM micro-organisms and used widely in the production of cheeses, yoghurts and other foods. A full list of products that have to be labelled is given in Table 4.2.

The regulations covering refined products are difficult to enforce because the products are identical whether they come from a GM or non-GM source and without the presence of DNA or protein there is no way of confirming a GM origin. It seems odd that a product that is identical to its non-GM counterpart but comes from a GM plant source should be labelled while there is no requirement to label a food that actually contains a modified enzyme from a GM micro-organism.

Although the stringent labelling laws undoubtedly made GM crops less attractive for European farmers and the food supply

Table 4.2 GMOs and GM foods that do or do not have to be labelled under EU regulations 1829/2003 and 1830/2003 that came into force in April 2004.

GMO or GM food	Labelling required?
GM plant or seed	Yes
Whole foods derived from a GM plant source	Yes
Refined products such as oil and sugar from a GM plant source	Yes
Foods containing GM micro-organisms	Yes
Animal feed containing GM crop products, including derivatives such as oil and gluten	Yes
Fermentation products produced for food or feed use	Yes
Food from animals fed GM animal feed	No
Food made with an enzyme produced in a GM micro-organism (some cheeses, for example)	No
Products containing GM enzymes where the enzyme is acting as an additive or performing a technical function	Yes
Flavourings and other food and feed additives produced from GMOs	Yes
Food containing GM ingredients that are sold in catering establishments	Open to interpretation. FSA opinion is yes.

(*Source*: Food Standards Agency.)

chain, and continue to do so, they may have had the effect of forcing some sectors to face the reality that major commodity-exporting countries have adopted GM crops widely. In 2007, for example, the Food Standards Agency compiled data on the global production of GM and non-GM soybean, maize, cotton and oilseed rape (Table 4.3), showing the penetration of GM varieties into the market. Although no official data are available, traded commodities are believed to be dominated by GM varieties to an even greater extent, with over 90% of traded soybeans likely to be GM and at least 50% of traded cotton and maize. These crops are indispensable for animal feed production and millions of tonnes

Table 4.3 Global production of four crops that are indispensable for animal feed production and the amount that is produced from GM varieties. (Figures compiled by the Food Standards Agency of the United Kingdom for 2007.)

	Total (tonnes)	GM varieties (tonnes)
Soybean	86 million	58.6 million (68%)
Maize	140 million	35.2 million (25%)
Cotton	34 million	15 million (44%)
Oilseed rape	23 million	5.5 million (24%)
Total for the above 4 crops	283 million	114.3 million (40%)

of grain and meal are imported into the UK for feed production every year. The implication of this is that it has become extremely difficult to source enough non-GM product to meet demand at an acceptable cost. This same conclusion has already led to feed for pigs and cattle being made with GM ingredients.

The poultry and fish feed sectors of the animal feed market still avoid the use of GM crop products, and the high end of the market also stipulates non-GM feed for cattle and pigs. This reflects pressure coming back up the chain from retailers, although it is not clear why it is considered important that poultry and fish are not fed with feed containing GM crop products while most cattle and pigs are. The fish feed market is an interesting one because consumers eat fish partly for omega-3 fatty acids. As described in Chapter 3, the long chain omega-3 fatty acids that make fish oil so valuable are made in marine algae and accumulate through the marine food chain; they are not made by fish themselves. Farmed fish, therefore, must have long chain omega-3 fatty acids in their diet if they are to accumulate them. This is currently provided in marine fish meal, making farmed fish no more sustainable as a source of long chain omega-3 fatty acids than marine fish, stocks of which are declining. An alternative and sustainable source of long chain omega-3 fatty acids should become available in the next few years in the form of GM soybean and oilseed rape, and it seems likely that the fish-farming industry will have to use it.

4.6 Safety Assessment and Labelling Requirements in the USA

New GM crop varieties have to undergo field trials in the USA in the same way that they do in Europe. However, much less detail is required in the risk assessment of the variety before the trial can go ahead. In fact, the procedure has been relaxed as more field trials have been undertaken and no problems have ensued, although this may be reversed as applications are made to test crops that produce pharmaceuticals or have other non-food uses. The total number of field trials of GM crops that have been run in the USA now runs into the tens of thousands, covering a variety of crop species and traits.

Test and commercial releases of GM crops in the USA are controlled by the Animal and Plant Health Inspection Service (APHIS) within the United States Department of Agriculture (USDA). Commercial use of GM crops that produce their own insecticide, such as Bt, is controlled by the Environmental Protection Agency (EPA). The safety of foods derived from GM crops is assessed by the Food and Drug Administration (FDA). In fact, there is no legal requirement for companies to seek the approval of the FDA but all of them do. Companies ask the FDA for an Advisory Opinion either on a specific characteristic of their product or on its suitability as a food. Liability for problems that arise after release rests with the company that markets the crop.

The first GM plant product to come onto the market in the USA was Calgene's Flavr Savr tomato (see Chapter 3). It underwent more than four years of comprehensive pre-market tests, the results of which were submitted to both the FDA and the USDA and published for public comment. Calgene requested two separate Advisory Opinions from the FDA, one on the use of the kanamycin resistance marker gene, the other on the status of the Flavr Savr tomato as a food.

The FDA issued a preliminary report that all relevant safety questions about the Flavr Savr tomato had been resolved. This was ratified by its Food Advisory Committee in a public meeting in April 1994. The FDA announced its findings that the Flavr Savr tomato was as safe as tomatoes bred by conventional means in May 1994 and Calgene began to market the new product shortly after.

The labelling laws in the USA are quite different from Europe. Products derived from GM plants do not have to be labelled as such. However, products that are significantly different from their conventional counterpart, such as high oleic acid soybean oil (see Chapter 3), do have to be labelled. In other words products are labelled according to their properties, not how they were made.

5 ISSUES THAT HAVE ARISEN IN THE GM CROP AND FOOD DEBATE

The controversy over GM crops has spanned the globe and lasted for a decade and a half with no sign of abating. It is arguably one of the great debates of our time and a testing ground for the ability of science and scientists to overcome irrationalism and scaremongering. In Europe there is no sign of the battle being won; indeed, so far it has not even been close. As a result, great damage has been done to the European plant biotechnology industry: Syngenta, Monsanto, Aventis (now part of Bayer), DuPont and Unilever have all closed or down-sized plant biotechnology operations in Europe in the last decade. Researchers have also been put at a great disadvantage. If a scientist in the USA wishes to undertake a field trial of a genetically modified crop he/she can submit a one-page application online and receive approval the next day. In Europe, scientists have to negotiate a complicated and expensive regulatory process (Chapter 4) and then face a high probability of their field trial being vandalised. Not surprisingly, only a handful of field trials have been undertaken in Europe in recent years.

The influence of European attitudes to GM crops extends well beyond Europe's borders. Europe is nowhere near self-sufficient in food and is a major market for commodity exporters. The European market is so lucrative that the countries that sell into it are extremely wary of anything that might jeopardise their ability to do so. This raises ethical issues: Golden Rice (Chapter 3), for example, which potentially could benefit millions of people in developing countries who suffer from vitamin A deficiency, has been held up for at least ten years by what its inventor, Ingo Potrykus, describes as 'European Attitudes'. It must be said that Europe is not entirely alone: Japan, for example, has been implacably opposed to GM crops and preserving its market in Japan has been the major reason for China not allowing the cultivation of GM rice until now (China approved GM insect-resistant rice varieties for the first time in 2010, although it could be 2012 before seed is made available to farmers).

Exactly why European consumers have been so much more fearful of GM crops than other consumers is not a simple question to answer. Opinion polls taken in Asian countries (other than Japan) and in the Americas consistently show that two-thirds or more of the population broadly approve of plant biotechnology and believe that they are likely to benefit personally from its development. Polls in the UK and Europe have consistently shown much less favourable attitudes amongst consumers. It should be noted that poll results depend in part on what question is asked, and the results of the United Kingdom's Food Standards Agency surveys on consumer attitudes (Figure 5.1a and b) show less hostility to GM foods amongst UK consumers than might be expected, given the tone of the debate that has taken place in the UK. One possible interpretation of this is that a small minority of people that are extremely hostile to GM foods have dominated the debate. The data also show that concern about GM foods is declining slowly amongst UK consumers (Figure 5.1a), that concern is expressed by only 5% of respondents if they are left to respond spontaneously, compared with 24% who select GM foods as an issue of concern from a list (Figure 5.1a), and that other issues concern consumers more than GM foods, with food poisoning top of the list (Figure 5.1b).

Europeans are not naturally 'technophobic'. Technology continues to develop rapidly on many fronts, from telecommunications to

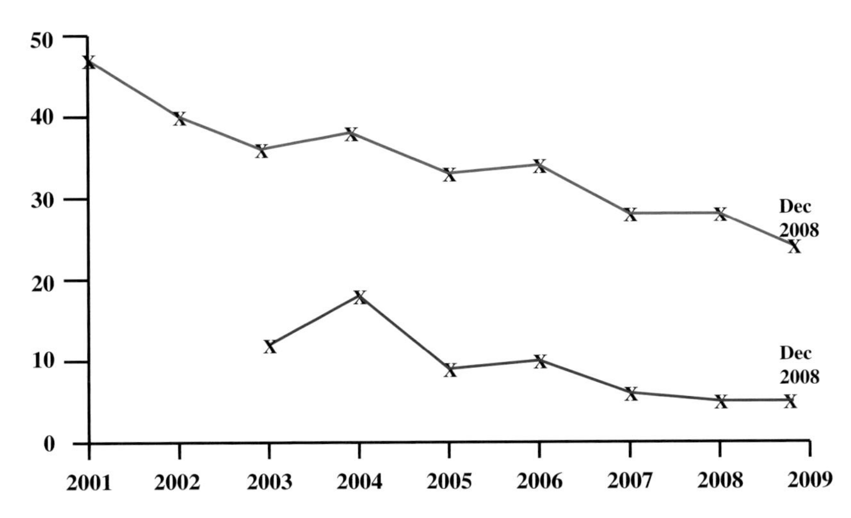

Figure 5.1a Results of United Kingdom's Food Standards Agency survey of consumer attitudes to GM foods from March 2001 to December 2008. All data points are for March in the year shown apart from the last one. The data on the red line shows the proportion (%) of consumers who select GM foods when asked 'Are you concerned about any of the following food issues' and provided with a list that includes GM. The blue line shows the proportion of consumers who cite GM foods when asked 'What food issues are you concerned about' but are not provided with a list. Data: Food Standards Agency.

computing and medicine and more, and most developments have been embraced by Europeans. The internet is just one example of a development that has had a far greater impact on society than GM crops, and people in Europe, as elsewhere, have accepted it, taking proportionate steps to minimise risks where they see them, for example by using parental controls to prevent their children from accessing inappropriate websites. Most Europeans also have a mobile phone, and that market grew astonishingly fast despite concerns being raised by some scientists about their safety. Another development that might have been expected to cause a heated debate was experimentation on human embryos, but while that was considered carefully by the science community and government there appeared to be relatively little interest amongst the general public. In contrast, experimentation on human embryos has been a

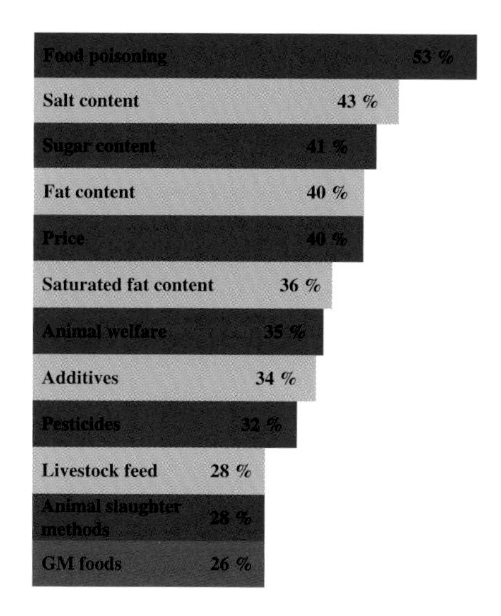

Figure 5.1b Results of United Kingdom's Food Standards Agency survey, December 2008, showing main food issues for consumers; respondents were given a list to choose from.

hugely controversial issue in the USA, while attempts by pressure groups to ignite a debate on GM crops have largely failed.

Part of the answer lies in the reluctance of Europeans to trust their governments or scientific experts on food issues. GM foods were launched in Europe shortly after the epidemic of bovine spongiform encephalopathy (BSE) in the UK cattle herd had led to one of the biggest food scares in UK history. Rightly or wrongly, consumers felt that they had been given the wrong advice by scientists and government ministers on the safety of beef. Food 'scares' are not unique to the UK and Europe, but the BSE issue did undermine the confidence of European consumers in a way that has not occurred elsewhere. This enabled pressure groups to dominate the debate, and European consumers have been bombarded with inaccurate information, half-truths and wild scare-stories, as well as bizarre and frightening imagery (Figure 5.2). Even if consumers do not believe the more hysterical of these stories, why should they take the risk of buying GM food products?

Figure 5.2 Images from the GM debate. Clockwise from top left: Joan Ruddock, Member of Parliament and head of the 'Five-year Freeze' campaign, behind a block of ice containing fish heads attached to corn cobs; campaign leaflet from Greenpeace, October 1997; billboard image used by 'Mothers Against Genetic Engineering' in New Zealand in a campaign to prevent a field trial of GM herbicide-tolerant onions; online advert from Marks and Spencers in 2008 arguably implying that other retailers might be selling GM bread (no GM wheat was or is grown anywhere); Prince Charles at a Soil Association press conference at which he stated that 'Mixing genetic material from species that cannot breed naturally, takes us into areas that should be left to God'.

Journalists have generally been happy to repeat the proclamations of pressure groups. Some of the wilder headlines from the UK include:

- 'Scientists warn of GM crops link to Meningitis', *Daily Mail,* 26 April 1999. The article continued, 'The nightmare possibility of GM food experiments producing untreatable killer diseases has been underscored by senior government scientists'. The story was retracted the following day in a short statement on an inside page.

- 'Are we at risk from mutant make-up?', *Express on Sunday*, 21 February 1999.
- 'Scientists raise the fear of GM foods triggering new allergies', *Express*, 30 April 1999.
- 'Lifting the lid on the horror of GM foods', *Express*, 12 May 1999.
- 'The GM pollen that can mean a cloud of death for butterflies', *Daily Mail*, 20 May 1999.
- 'Mutant porkies on the menu', *News of the World*, 23 May 1999.
- 'GM risk in daily food of millions', *Guardian*, 24 May 1999.
- 'GM food "threatens the planet"', *Observer*, 20 June 1999.
- 'Meat may be tainted by Frankenstein food', *Daily Mail*, 6 July 1999.
- 'M&S sells genetically modified Frankenpants', *Independent on Sunday*, 18 July 1999.
- 'Is GM the new thalidomide?', *Daily Mail*, 8 October 2003.
- 'How GM crop trials were rigged', *Independent on Sunday*, 12 October 2003.
- 'Stop the rush to GM crops', *Independent on Sunday*, 12 October 2003.
- 'Curb on GM crop trials after insect pollution', *Daily Telegraph*, 14 October 2003.
- 'Polluted for generations', *Daily Mail*, 14 October 2003.

Scientists in the UK have been criticised for losing the GM debate, through failing to engage with the public or the media, failing to communicate in terms that non-scientists can understand, or reacting too slowly to issues when they arose. However, it could be argued that there is not a lot that scientists could have done in the face of a campaign that was waged at the level demonstrated by the headlines above. In any case, few scientists are trained to deal with the media or the public and most find it a difficult and uncomfortable experience. A 'correct' answer to the question 'Are GM foods safe?' might be: There is an international consensus amongst experts in the field that GM is not inherently more risky than other methods in plant breeding; GM foods are examined under a rigorous assessment system that goes beyond that applied to other foods; the safety of GM food has been considered by the World Health Organisation, the European Union, the Organisation for Economic Co-operation and Development, as well as many national governments, and none has concluded that there is any evidence of adverse effects on health. However, there is little chance of getting through the first line of this statement before a television interviewer interrupts or the audience starts thinking about something else. The opposing statement that GM foods are unsafe or untested may not be true but it is much simpler to put across. So should scientists simply answer yes to questions such as 'Are GM foods safe'? Perhaps a decade and a half of safe use of GM crops and consumption of GM foods by hundreds of millions of consumers without the report of so much as a bellyache justifies a 'Yes' answer, but it is a difficult ethical issue for scientists.

So what are the issues and concerns that have been raised during the GM crop debate? The rest of this chapter outlines and attempts to address some of them.

5.1 Are GM Foods Safe?

As discussed above, scientists are always reluctant to describe anything as completely safe. However, if we accept that new crop varieties produced by non-GM methods are safe enough, it is reasonable to judge the safety of GM crops in comparison. The major arable and horticultural crops grown and consumed in western Europe have been developed over centuries or even millennia. Consequently, they are assumed by the consumer to be safe and wholesome. However, most, if not all, of these crops contain compounds that are potentially toxic or allergenic. In most cases, these compounds have probably evolved to provide protection against animal grazers or pathogenic micro-organisms and it is therefore not surprising that they are also toxic to humans. They include glycoalkaloids in potatoes, cyanogenic glycosides in linseed, glucosinolates in *Brassica* oilseeds and proteinase inhibitors in soybean and other legume seeds. Some of these have been discussed previously in this book.

It is very doubtful whether these or many other generally accepted foods would be approved for food use were the toxins introduced by genetic modification. Nevertheless, with some exceptions, the introduction of new types and varieties of food crops produced by conventional breeding requires no specific testing for the presence of allergens or toxins, even if genes have been introduced from exotic varieties or related wild species by wide crossing.

The products of genes introduced by genetic modification are readily identified in a GM plant. They may also be isolated in a pure form, either from the species of origin, from the GM plant or after expression in a micro-organism. The pure protein can then be tested in detail and its presence in processed foods monitored. In contrast, it is virtually impossible to predict or characterise all of the changes in food composition that may result from conventional plant breeding. As for unpredictable effects, as described in Chapter 4, environmental factors have been shown to have more effect on the composition of wheat grain, for example, than genetic modification, while gene expression differs much more between different genotypes produced by conventional breeding than between different GM genotypes.

On top of all this, as described in Chapter 4, GM crops and foods are examined under a rigorous assessment system. Consequently, it would not be unreasonable to argue that GM foods may be safer than food derived from new, non-GM varieties.

5.2 Will Genetic Modification Produce New Food Allergens?

It has been suggested that consumption of GM foods could lead to increases in allergenicity. In fact, there has already been a widely reported case of increased allergenicity in a GM plant line, but the problem was detected and the product never reached the market. A methionine-rich 2S albumin storage protein gene from Brazil nut was expressed in soybean in order to increase the content of methionine, an amino acid, in soybeans used for animal feed. The protein was subsequently shown to be an allergen, as are a number of related 2S albumins from other species, and the programme was discontinued.

As with the more general question of safety, it could be argued that genetic modification is no more or less likely to lead to an increase in food allergens than other methods in plant breeding. Nevertheless, the industry and regulators are clear that every precaution should be taken to ensure that no new allergens are introduced into the food chain through genetic modification.

In fact, the biotechnology industry is bullish about its ability to predict whether or not a GM food is likely to be more allergenic than its traditional counterpart and to detect new allergens before a product reaches the market if they get their predictions wrong. A typical procedure described by Astwood and co-workers combining molecular analyses of the novel gene and protein, immunoassays and skin prick tests is shown in Figure 5.3.

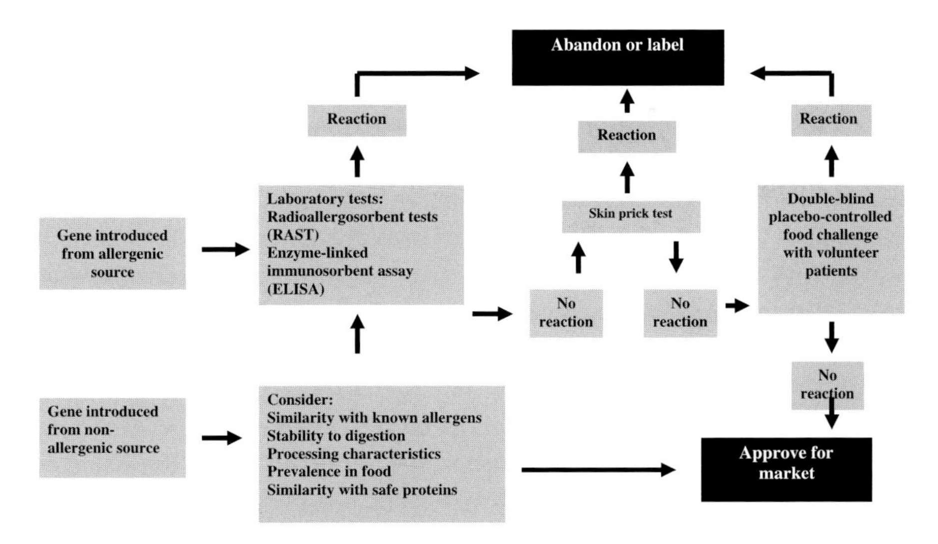

Figure 5.3 Decision tree showing the assessment and testing of possible allergenicity in GM foods, as described by Astwood *et al.* (*Food Allergy*, 2nd ed., Blackwell Science, 1996).

5.3 Is it Ethical to Transfer Genes Between Different Species?

This question has been raised famously by Prince Charles. He has been quoted as saying that 'Mixing genetic material from species that cannot breed naturally, takes us into areas that should be left to God'. Of course, plant breeders were mixing genetic material from sexually incompatible species long before genetic modification was developed (see the section in Chapter 1 on triticale as an example). Nevertheless, this statement encapsulates the views of a significant number of people. It is a pity that the point could not have been made without the use of overblown language, such as the term 'Frankenstein Foods'.

The philosophical counter-argument to this view is that all life on Earth has a common ancestor if you go back far enough and that the species alive today are part of an evolutionary continuum, not separate entities. So genes encoding proteins involved in cellular control mechanisms that evolved long ago, such as the control of cell division, for example, are closely related and instantly recognisable to those that work on them, whether they come from yeast, invertebrates, plants or man. So how can it be fundamentally wrong to move genes between species?

The pragmatic counter-argument is that it is not where a gene comes from or how it is brought into a plant breeding programme that is important but what it does. A gene producing a poison that is produced inadvertently by radiation mutagenesis or crossed into a crop plant from a wild relative is clearly much more dangerous than a gene producing a benign protein that is introduced into a crop plant by genetic modification. Similarly, genes imparting herbicide tolerance can 'flow' into weed plants by cross-pollination regardless of whether they have been engineered or crossed into a crop plant or produced by radiation mutagenesis.

This argument, which has a sound scientific basis, would lead to the conclusion that GM crop varieties should be assessed and regulated no differently to new crop varieties produced by other means, but that is not the case. For example, as described in

Chapter 3, GM oilseed rape varieties that are tolerant to one or other of two broad-range herbicides, glyphosate and gluphosinate, are grown widely in North America. There is another broad-range herbicide-tolerance trait available in commercial oilseed rape varieties in Canada, in this case the herbicide being imidazolinone. This trait was developed by Pioneer Hi-Bred, now part of DuPont, but it was produced by mutagenesis, not genetic modification. Oil or meal produced from this variety can be imported into Europe without having to pass through the regulatory system that the GM varieties have had to go through, and without any of the safety assessments. Furthermore, although imidazolinone is not used in Europe, in theory there is nothing to stop a plant breeder crossing this trait into a European variety and marketing it in Europe, again without having to go through the regulatory system that so hinders the development of GM varieties. It is difficult, therefore, to argue that the system is science-based.

Part of the reaction against moving genes between species is an emotional one. The use of an 'antifreeze' gene from a fish to engineer frost resistance into tomatoes, for example, caused a much greater reaction than the use of a bacterial gene to engineer insect resistance into maize, despite the fact that the tomatoes were not a commercial product whereas the maize was. So far, biotechnology companies have avoided using animal genes in GM crops for food use but are using them to make pharmaceutical products in GM plants. Clearly, the industry believes that consumers have different attitudes when they are buying and using pharmaceuticals to when they are buying and consuming food.

5.4 Animal Studies

The World Health Organisation is sceptical about the usefulness of animal feeding studies in the safety assessment of GM plants and foods. This scepticism arose in part from experience gained in the testing of irradiated food in the early 1990s. Animal studies are undoubtedly useful in the toxicological safety assessment of individual compounds such as pesticides, pharmaceuticals, industrial chemicals and food additives. It is relatively simple to feed such compounds to animals at doses sometimes far higher than humans would be exposed to and to identify any potential adverse effects on health. Foods, on the other hand, are complex mixtures of compounds. Animals cannot be persuaded to eat orders of magnitude more of them than humans can, and feeding only one type of food to an animal usually reduces the nutritional value of the diet, causing adverse effects that are not related directly to the material itself. For these reasons, relating any effects on the welfare of an animal to a particular genetic modification can be extremely difficult.

Nevertheless, the World Health Organisation recommends the use of animal studies if other information available on the GM plant or food is inadequate, particularly if the novel protein in the GM food has not been present in the food chain before. It also supports the use of animal testing of proteins produced by novel genes before the gene is used in biotechnology (in other words, for the gene to be engineered into a micro-organism so that large amounts of its protein product can be made and purified; the individual protein can then be used in toxicology studies).

The fact that GM foods did not have to be tested on animals was used by pressure groups in the late 1990s and early 2000s to undermine consumer confidence in GM food safety. Partly for this reason, all applications made in Europe for consent to market foods produced from GM crops now include the results of animal feeding experiments, despite the ethical question of whether it is right to undertake studies on animals if the results are unlikely to be meaningful.

5.5 GM Crops 'Do Not Work'

A long-running strategy of anti-GM campaigners is to claim that GM crops are flawed in some way. In other words, there is a problem inherent in the technology that means that no GM crop will deliver what it promises. GM crops were linked by activists with crop failures in India and Indonesia in the 1990s, for example, despite the fact that neither of these countries was using GM crops at that time, and with farm failures in the USA, where farmers continue to use GM crops in increasing numbers.

In its report, 'Seeds of Doubt', the Soil Association, a UK organic farming group, claimed that the use of GM crops in the USA had reduced profitability, reduced yields and raised costs through increased herbicide use. Herbicide-tolerant soybean was cited as an example of a crop that had failed to meet expectations of higher yields. In contrast, a much more extensive report from the National Centre for Food and Agricultural Policy cited herbicide-tolerant soybeans as the most successful GM crop in improving farm profitability. Increased profits are gained through reduced costs, not improved yield. Indeed, improved yield was never a target for herbicide-tolerant crops in a developed country where farmers have access to a variety of weed control measures. So why do the two reports differ? Perhaps because the Soil Association report, as it states itself, did not cover positive experiences.

Not all new crop varieties, GM or otherwise, are successful, and some GM crop failures are described in Chapter 3. Indeed, one of the first GM varieties on the market, the 'Flavr Savr' tomato, was a commercial failure and was soon withdrawn. However, some GM crop varieties have been extremely successful. In 1996, when GM herbicide-tolerant soybean was first grown widely, few people even in the biotechnology industry would have predicted that by 2009 more than three-quarters of the world's soybean crop would be GM. It is not the purpose of this book to 'push' the use of GM crop varieties in favour of others. However, it is appropriate to point out that the major GM crop varieties have now been available in the USA and elsewhere around the world for 15 years, long enough for farmers to make a judgement. The continued popularity of these varieties indicates that farmers find some benefit in using them.

5.6 Did Tryptophan Produced by Genetic Modification Kill People?

This issue was often brought up when GM crops first came onto the market, although it concerns GM bacteria, not plants.

Tryptophan is an amino acid and is essential to all living things in order for them to make proteins. Plants, fungi and bacteria make their own but animals, including humans, do not and need to acquire tryptophan in their diet. Experiments with rats fed tryptophan-deficient diets suggest that tryptophan deficiency delays growth, development and maturation of the central nervous system. As well as its role as an amino acid required for protein synthesis, tryptophan is a precursor for a neurotransmitter, serotonin. There is no evidence that tryptophan deficiency is a problem in developed countries since it is plentiful in meat and dairy products and a variety of plant sources, but health supplement manufacturers claim an extensive range of benefits associated with taking tryptophan as a supplement.

Many pharmaceutical companies around the world therefore manufacture tryptophan, usually from selected strains of bacteria. One of these, Showa Denko in Japan, used a GM strain of *Bacillus amyloliquefaciens* to produce the raw product in the late 1980s. In 1989 the company changed the strain of bacterium that it was using and reduced the number of purification processes that it used to clean up the raw product. The result was that certainly hundreds, probably thousands of people in the USA who bought the product under various brand names became ill with eosinophilia-myalgia syndrome (EMS). Tragically, 38 people died. Tryptophan was withdrawn as an over-the-counter product in the USA and the Food and Drug Administration has never allowed its return, although it can be obtained on prescription.

An investigation into the incident found that several contaminants that should have been removed in the purification process were present in the finished product. Why they were still present is not clear since Showa Denko did not release the results of its own internal inquiry. It is also still not certain which contaminant was responsible for the illnesses and deaths. No manufacturer is currently using GM bacteria to make tryptophan, but problems with contaminants and the resulting condition of EMS still occur sporadically, which suggests that the problem is one of old-fashioned chemical engineering, not genetic modification.

The irony of this story for plant biotechnologists is that although it concerns GM bacteria, it was used as an argument against the genetic modification of plants. Meanwhile, the use of GM bacteria to produce medicines, vaccines, food additives and supplements attracts little attention.

5.7 The Monarch Butterfly

In 1999 a study conducted by John Losey and his team at Cornell University found that caterpillars of the monarch butterfly that were forced under laboratory conditions to eat large quantities of pollen from GM insect-resistant maize suffered higher mortality levels than caterpillars that were not fed the pollen. The study was published in the journal *Nature* and caused an international furore. Indeed, because of the iconic status of the monarch butterfly in the USA, there were fears that the study might cause the sort of public backlash against GM crops there that had occurred in Europe. As it turned out, although the issue was more grist to the mill for the media and anti-GM activists in Europe, Americans remained broadly supportive of plant biotechnology.

The *Cry1A*, or Bt gene (see Chapter 3) that is engineered into maize to make it insect-resistant is toxic to caterpillars. The gene is designed to be active everywhere in the plant, including the pollen (subsequent varieties may contain genes that are designed only to be active where they are needed), so it is no surprise that Monarch butterfly caterpillars that ate the pollen did not thrive. However, this was a laboratory experiment. Monarch butterfly larvae eat milkweed, not maize pollen (in the experiment the pollen was spooned onto milkweed leaves so that the larvae had no choice but to eat it). Even as the first report was published, experts were extremely sceptical that maize pollen would ever accumulate in such amounts on milkweed in the wild. Field-based studies published subsequently bore this out.

Similar laboratory-based experiments have shown that the survival rate of predator species such as lacewings and ladybirds can be reduced if they are fed exclusively on prey species that are feeding on GM insect-resistant plants. None of these results have been replicated in the field. It should be remembered that spraying caterpillars, lacewings, ladybirds and other insects with pesticide, which is the practice for non-GM maize, kills them all outright. The use of GM insect-resistant plants in the field can lead to an increase in beneficial insects. With GM maize, for example, the farmer does not have to use an early spray against the corn borer. The farmer may need to spray against other pests but, if this can be avoided, benign predatory insects can thrive. If this happens, the predatory insects prevent a late infestation of red spider mites, reducing pesticide use even further.

As soon as the initial report was published in 1999 and the controversy erupted, Losey was at pains to point out that the study was a preliminary one and that field studies would be required to find out the real impact of GM crops on butterfly populations. This raised eyebrows amongst the science community because *Nature* does not normally publish preliminary reports. Indeed, many scientists felt that the study would not have been considered worthy of publication without the GM crop tag and that publication was rushed through in order to generate headlines and publicity.

5.8 The Pusztai Affair

The other incident that brought the GM issue to the fore in the late 1990s involved Dr Árpád Pusztai, who was working at the Rowett Institute in Scotland. Dr Pusztai had produced GM potatoes that were engineered to produce a type of lectin. Many plants produce lectins as natural insecticides but most lectins are unsuitable for use in plant biotechnology because they are poisonous to animals, including humans. The lectin gene that Pusztai used came from snowdrop. When the GM potatoes were fed to rats, the rats suffered from lectin poisoning, which is, perhaps, not very surprising. However, Pusztai claimed to have fed another group of rats with ordinary potatoes that had been spiked with the same lectin. These rats, he said, suffered from lectin poisoning but in his view they did not suffer as badly as those fed with the GM potatoes. Pusztai therefore claimed that the process of genetic modification had somehow made the potatoes more poisonous than they would have been otherwise.

The normal method of dissemination of scientific results is through the scientific press. Every article is reviewed by at least two independent experts in the field, the aim being to make sure that experiments have been conducted properly and that the deductions made by the authors of the article are justified by their data. This system is not infallible but no-one has come up with a better one. However, on 10 August 1998, before the study had been submitted for peer review, a documentary in the *World in Action* series was broadcast on national television in the UK in which Dr Pusztai claimed that the effect of feeding GM potatoes to rats 'was slight growth retardation and an effect on the immune system'. On the same day, a Rowett Research Institute press release made the same claims and called for further research.

There was huge, international media interest in the claims and this was heightened even further when, two days later, the Rowett Institute published another press release announcing that Dr Pusztai had been suspended and would be retiring. It turned out that some of the experiments that had been referred to in the *World in Action* programme had not been carried out, but the announcement of Dr Pusztai's suspension and retirement led to speculation in the media that his evidence of potential dangers of GM foods was being suppressed.

The affair became such an important public issue that the experiments and results were reviewed by a group of scientists appointed by the Royal Society, who concluded that the work was 'flawed in many aspects of design, execution and analysis' and that 'no conclusions should be drawn from it'. Nevertheless, a medical journal, *The Lancet*, published the study in 1999, giving it a degree of credibility that it did not deserve.

The incident was considered by the UK's Parliamentary Select Committee on Science and Technology in its inquiry into genetic modification. The report of the inquiry is available at: www.parliament.the-stationery-office.co.uk/pa/cm199899/cmselect/cmsctech/286/28605.htm

Dr Pusztai admitted to the inquiry (paragraph 26 of the report) that 'no differences between parent and GM potatoes could be found', directly contradicting his statement in the *World in Action* programme. Nevertheless, as the report goes on to state, the press continued to give credibility to Dr Pusztai's claim, despite it being contradicted by his own evidence, and still does. Pusztai became a hero of anti-GM campaigners, to whom he acted as a consultant long after his retirement.

5.9 Alarm Caused by Contradictory Results of Biosafety Studies

The Pusztai study is not the only one to have produced results contradicting the scientific consensus on GM crop and food safety and to have caused great controversy after being publicised without scientific peer review. Herbicide-tolerant soybeans, for example, were approved for food use in the 1990s and have been widely eaten and studied since then, with no alarm being raised. However, in October 2005, Dr Irina Ermakova of the Institute of Higher Nervous Activity and Neurophysiology of the Russian Academy of Sciences reported that she had conducted a study that showed much higher death rates and lower body weight in young rats whose mothers were fed a diet of herbicide-tolerant GM soybeans compared with rats fed non-GM soybeans. The results were reported without being scientifically peer-reviewed.

Toxicology experts, including those on the relevant UK expert advisory committees and the European Food Safety Authority, complained that they were unable to perform a serious review of the study because insufficient information was available on how the experiments had been conducted. The UK's Advisory Committee on Novel Foods and Processes (ACNFP) stated that they considered the results sketchy and inadequately supported, with no information on the composition of the rats' diets. They also pointed out that the study's findings contradicted peer-reviewed scientific studies that had not shown any negative health effects. For example, Denise Brake and Donald Evenson at South Dakota State University had conducted similar feeding studies on mice and published their results in a peer-reviewed scientific journal, *Food and Chemical Toxicology*, in 2004. They had found no negative effect of the GM soybeans. That did not stop the Russian newspaper *Pravda* predicting that life expectancy would decrease as a result of the widespread consumption of GM soybeans, or the UK's *Daily Mail* warning of risks posed to unborn babies.

GM maize has also come under attack, again despite being approved and used in food and feed since the mid-1990s. In 2007, a French scientist, Gilles-Eric Séralini of the University of Caen, questioned the interpretation of the results of animal feeding experiments with GM insect-resistant maize. The French government used this as an excuse to stop the growing of GM maize in France, in defiance of the European Commission. GM maize cultivation had been gaining popularity with farmers in France in 2006 and 2007 (Figure 3.3) but had become increasingly controversial.

In 2008, the Austrian government also released the results of a study, without peer review, from Jürgen Zentek at the University of Vienna that claimed to show fertility problems in mice fed a particular variety of GM maize. The publication of the report caused considerable controversy and was used by Austria and other countries to justify their bans on GM crops and foods. However, the Austrian government announced in October 2009 that they were withdrawing the study, saying that Zentek and his co-workers had failed to deliver a satisfactory report on it, especially with respect to the statistical analysis of the data, and that the government no longer expected to receive such a report. In April 2010, EFSA reported on its re-evaluation of the GM maize varieties that were approved in Europe, confirming that they considered there to be no scientifically based cause for concern.

5.10 'Superweeds'

The term 'superweed' is an example of the emotive hyperbole that has dominated the debate on GM crops in Europe. It arises from the notion that a gene that imparts herbicide tolerance in a GM crop could 'flow' into related wild species through cross-pollination and the resulting hybrids would become uncontrollable weeds.

The potential environmental impact of cross-pollination of GM crops with wild species has to be assessed case by case, taking into account the species and genes involved. Maize and potato, for example, do not cross with any wild species in the United Kingdom (although forced crosses can be made between potato and black nightshade in the laboratory) and wheat does not cross with any native plant species to produce fertile hybrids. Wheat is also almost entirely self-pollinating. Oilseed rape, on the other hand, will cross with other cultivated and wild *Brassicas*, including Chinese cabbage, Brussels sprouts, Indian mustard, hoary mustard, wild radish and charlock. The extent of such crossings in agricultural systems is the subject of continuing research, but it does not necessarily mean that GM oilseed rape represents a problem, just that the potential risk represented by GM oilseed rape is different to that of wheat. Recent studies carried out in Canada, where there are millions of hectares of GM herbicide-tolerant oilseed rape, have not found any hybrids between the GM crop and wild relatives.

Even if a hybrid between a GM crop plant and a wild relative did arise, it would only prosper if the gene involved could confer a competitive advantage. Herbicides do not exist outside agriculture, so herbicide-tolerance genes would be unlikely to persist in the wild. Weed populations that acquire herbicide tolerance genes from crop plants or that evolve tolerance themselves could become a problem for farmers and if the problem were to become too serious the herbicide and the GM crop that went with it would become useless. Farmers would have to turn to other varieties and different herbicides. Farmers have been aware of these problems since herbicides became used widely, long before genetic modification was developed. Farmers will also have to be careful that GM crops with different herbicide-tolerant traits do not cross to produce multi-tolerant hybrids.

5.11 Insect Resistance to Bt Crops

Just as the potential for weeds to develop resistance to control measures is never far from the minds of farmers and those involved in agrochemical production, so is the potential for insect pests to develop resistance to pesticides. This pre-dates genetic modification but the introduction of GM plants containing their own insecticide brought a different aspect to this old problem. In the USA, the responsibility for monitoring and controlling GM plants that had been engineered to be insect-resistant through the introduction of a *Cry* gene (see Chapter 3) was given to the Environmental Protection Agency (EPA). The EPA was already responsible for controlling the use of the Bt pesticide and argued that since both the GM plants and the pesticide used the same insecticidal protein the two had to be controlled by the same agency. This makes sense in some ways but other GM crops are controlled by the Animal and Plant Health Information Service (APHIS) within the US Department of Agriculture and this splitting of responsibility for GM crops may have contributed to the StarLink fiasco (see Section 5.17).

The Bt pesticide is an important one for organic and salad farmers in the USA and elsewhere. Organic farmers because the Bt pesticide is allowed by most organic accreditors, and salad farmers because Bt is considered safe enough to allow its use right up to the point of harvest. However, the emergence of resistance to the pesticide was considered to be a significant risk and this had been a long-term concern for the EPA. In response to the introduction of Bt GM crops in 1996, the EPA insisted that farmers using GM crops containing the *Cry1A* gene would have to plant a proportion of non-GM crop as well. This would provide a refuge in which insects that had developed resistance to the effects of the Bt protein would not be at a selective advantage (in fact they would be at a selective disadvantage). According to the EPA's predictions of how the insect population would respond, the proportion of non-GM crop would have to vary according to what other insect-resistant GM crops were being grown in a particular area.

If the *Cry1A* gene were to 'flow' into weed plants through cross-pollination, the models on which the EPA bases its predictions of how insect populations are likely to respond would break down. For this reason the EPA proposed banning GM insect-resistant cotton in areas of the USA where cotton has wild relatives.

As is always likely in the USA, this proposed interference from a Federal Government Agency in Washington in what varieties farmers could or could not grow went down very badly in the Midwest. In response to criticism from farmers' representatives, the EPA published a report that stated that if its recommendations were not followed there would be widespread resistance to Bt in insect populations within three to five years. Both the insect-resistant GM crops and Bt pesticides would then be useless. The EPA got its way and the refuge policy was introduced. So far, 15 years later, it appears to have been very successful. However, activists in Europe lifted the worst-case scenario prediction out of the EPA's report and used it as one of the justifications of their opposition to GM crops, predicting the emergence of 'superbugs'.

5.12 Segregation of GM and non-GM Crops: Co-existence of GM and Organic Farming

Unless GM food is accepted universally, which seems unlikely in the foreseeable future in Europe, it is important that alternatives remain available to allow consumers to exercise choice. For imported food-stuffs, European suppliers will either have to buy produce from countries where GM crops are not grown, which will become increasingly difficult, or pay farmers overseas to grow non-GM varieties. This means paying a guaranteed price to a farmer to use old-fashioned varieties and high chemical inputs. So far, European buyers are apparently reluctant to commit to paying an increased price for non-GM soybean, with the exception of the UK.

If GM crops are to be grown widely in Europe, one of the key issues that will arise is that of segregation of GM and non-GM crops and food. Segregation could break down through accidental mixing of GM and non-GM seed for planting, through cross-pollination between GM and non-GM crops (not a problem for inbreeding species such as wheat) or through mixing of the product between the farm gate and the consumer. Some inadvertent mixing is almost inevitable and the production of totally GM-free food is therefore likely to be expensive and difficult.

The organic farming industry in Europe and the USA has attempted to corner the GM-free market. Indeed, presenting itself as GM-free has been an extremely successful marketing strategy even in countries such as the UK where other farmers are not using GM crops either. As part of this strategy, the organic industry persuaded authorities in the USA and Europe to write GM-free into the official definition of organic produce. This has led to the bizarre situation where an organic farmer can spray Bt pesticide onto a crop and still sell it as organic, but cannot grow a crop that has been engineered to produce the Bt protein itself.

Clearly, if organic and GM farming are to co-exist there will have to be tolerance on both sides. Organic farmers will have to accept that there might be some cross-pollination of their crops from GM varieties planted in adjacent fields. On the other hand, farmers who have paid for expensive GM seed in order to produce a high-value product will have to accept that some of the seed produced by the crop will result from cross-pollination from old-fashioned varieties grown by the organic farmer next door. Studies in Australia have suggested that a tolerance level of 1% (i.e. a crop with 1% GM content could still qualify as organic) would be sustainable even for out-breeding crops such as oilseed rape. It should also be remembered that there is considerable experience in the UK and elsewhere in Europe of segregating different crop varieties. A good example is the segregation of oilseed rape varieties suitable for industrial uses from those bred to be used for food (see Chapter 3), something that has been done successfully for several decades.

Organic farmers represent a tiny minority of the total, selling into a niche market with relatively affluent consumers. Almost no organic sugar beet or oilseed rape is grown in the UK, for example, because there is no market for it. While the interests of organic farmers are important, therefore, self-imposed rules on the incompatibility of GM crops with organic farming should not be allowed to affect the availability of GM varieties for farmers producing food for the wider public.

5.13 Antibiotic Resistance Marker Genes

Another topic that has generated a great deal of debate is the use of antibiotic resistance genes as selectable markers. The use of marker genes to select cells that have been modified with genes of interest is discussed in Chapter 2, and antibiotic resistance genes have been extremely valuable in the development of GM technology. Many scientific bodies around the world, including the World Health Organisation and regulatory committees set up by the European Union and several national governments, have considered the safety of antibiotic genes in food and have concluded that those that are being used do not represent a health threat. The British Medical Association, however, have expressed reservations and the United Kingdom's Advisory Committee on Novel Foods and Processes has called for the development of alternative marker systems. The biotechnology industry considers antibiotic resistance marker genes to be safe but a public relations liability and it is unlikely that they will be used in commercial GM varieties in the future. Nevertheless they will continue to be used in basic research.

The supposed risk associated with the use of antibiotic resistance marker genes is that they will somehow find their way into gut or soil bacteria and from there into disease-causing bacteria. In fact, the antibiotic resistance marker genes used in plant biotechnology confer resistance to antibiotics such as kanamycin and neomycin that are not used at all in oral medicines. Furthermore, these genes are designed to work in the plant and would not be active in bacteria.

Antibiotic resistance marker genes are also used in the genetic modification of bacteria, including bacteria used for the cloning, manipulation and bulking up of genes to be used in plant transformation. Where whole plasmids are used to transform plant cells by particle bombardment (see Chapter 2), such a gene may be incorporated into the plant genome along with the gene of interest. The antibiotic resistance marker genes used for the genetic modification of bacteria include one that imparts resistance to ampicillin, which is used in medicine. This gene was reported to be present in an early variety of GM insect-resistant maize but is not present in any of the varieties used today.

So is there any risk associated with the presence of such a gene in a GM plant? Firstly, the gene is integrated into the genome of the GM plant and the risk of horizontal gene transfer from plant genomes to soil or gut bacteria under natural conditions is extremely low, if not non-existent. Furthermore, where did these genes originate from in the first place? The ampicillin resistance gene came from a human gut bacterium, *Escherichia coli*. In fact, even a cursory examination of the information available shows that many natural bacterial species contain an ampicillin resistance gene, including *Kluyvera ascorbata*, *Pyrococcus furiosus*, *Proteus mirabilis*, *Bacillus subtilis*, *Klebsiella pneumoniae* strain H18, *Klebsiella pneumoniae* strain G122, *Pseudomonas aeruginosa*, *Pyrococcus furiosus*, *Staphylococcus aureus*, *Synechocystis* sp. PCC 6803, *Sinorhizobium meliloti*, *Yersinia pestis* strain CO92, *Mycobacterium leprae*, *Deinococcus radiodurans*, *Vibrio cholerae* and many soil bacteria. In other words the 'scare' over antibiotic resistance genes used in plant biotechnology is that they could somehow find their way into bacterial populations that already have them.

Antibiotic resistance marker genes are only maintained in bacterial populations if they impart a selective advantage. Antibiotic resistant strains of pathogenic bacteria do represent a health threat, but they arise naturally and thrive because of the sloppy management of antibiotics in human and animal medicine, not because of the use of antibiotic resistance marker genes in biotechnology.

5.14 Patenting

Patenting law was devised for two reasons: firstly to allow inventors to benefit from their inventions, secondly to bring inventions into the public domain so that they would not 'die' with their inventors. The invention must be described in sufficient detail to allow others to repeat it and must satisfy three criteria for the patent to be approved: it must be novel, it must be useful and it must be 'non-obvious' even to an expert in the field.

All of this seems quite laudable and the system has generally achieved its objectives. Patents protect inventions for up to 20 years, enabling inventors to prevent competitors from copying their invention or to license use of the technology at a price. Any new gadget is likely to be covered by a plethora of patents and licensing agreements.

Traditionally, patents were not awarded to new plant varieties or to strategies devised by plant breeders to improve crops. Indeed, until relatively recently plant breeders had almost no commercial protection at all for their new varieties. This changed in 1961 with the advent of Plant Breeders' Rights, a form of intellectual property designed specifically to protect new varieties of plants. Plant Breeder's Rights were drawn up at The International Convention for the Protection of New Varieties of Plants (the UPOV Convention) in Geneva in 1961 and became law in the United Kingdom in 1964. They were revised and strengthened in 1991, although the new regulations did not come into force in the UK until 1997.

Plant Breeder's Rights enable the holder of the rights to prevent anyone from producing, reproducing, offering for sale or other marketing, exporting, importing, conditioning for propagation or stocking a new variety without a licence from the holder. However, Plant Breeder's Rights do not prevent another breeder from using a variety in their own breeding programme.

Since genetic modification was developed, many patents have been filed covering the use of specific genes, types of genes, gene promoters, GM plant varieties and various technologies used to transform plants. Most of these patents have never been tested in court and it is still unclear to what extent they will be effective.

Many people in the industry are sceptical about their usefulness but continue to file patents if only to ensure that they are not shut out by someone else doing so.

More confusion is generated by the fact that different countries have different patenting laws and interpretations. European patent examiners have been very stringent in making sure that biotechnology patents satisfy the three criteria listed above while their American counterparts have been more inclined to accept a patent filing and let the courts sort it out later if the patent is challenged. Canada does not allow patents that cover plants or animals at all. The World Trade Organisation is attempting to uphold patents covering plant varieties worldwide under the General Agreement on Trade and Tariffs (GATT) but it has its work cut out.

The advantage for a company in holding a patent on the use of a particular gene in plant biotechnology is that the gene can then only be used under licence. Anyone who wants to incorporate the gene into their own breeding programmes or even farmers who want to grow a crop containing the gene can only do so with permission from the company holding the patent. The biotechnology company has the right to charge a royalty for use of the gene or of crops that contain the gene. It is therefore able to make money out of its technology in ways that Plant Breeder's Rights would not allow.

Patenting law does not allow for so-called 'bio-piracy', the raiding of a developing country's biological resources by a western company. A patent on a particular species of tree, or a traditional crop, would not satisfy the three criteria listed above and would be thrown out. However, it is possible to patent the idea of using a particular gene from any source to engineer a trait into a crop plant. Scientists in countries such as India and China are well aware of this and are keen to exploit the huge biological potential of their native flora, which dwarfs that available to scientists in temperate countries.

5.15 Loss of Genetic Diversity

It is widely believed that the introduction of GM crops will reduce biodiversity in agriculture, or more accurately it will reduce the genetic diversity available to plant breeders in the form of different crops and varieties. The great success of, for example, herbicide-tolerant soybean might at first sight appear to support this idea. However, the fact that a particular trait becomes popular with farmers does not necessarily mean that a single variety will be grown by everyone. The glyphosate-tolerant trait has been licensed by Monsanto to many other plant breeders who have crossed the original line with their own varieties to produce new varieties that carry the herbicide-tolerance gene but are suited for particular local growing conditions. Approximately 150 different seed companies now offer glyphosate-tolerant varieties of soybeans.

Plant breeders are acutely aware of the risks of too much inbreeding and the potential of wild and unusual genotypes for providing genes that might improve a crop breeding line. For this reason, they conserve old and wild varieties in case they can be used in a breeding programme. Potato breeders, for example, have access to over 200 cultivated potato lines and 8,000 wild potato relatives that have been characterised. This is more genetic diversity than they are ever likely to know what to do with.

5.16 The Dominance of Multinational Companies

Much of the basic science and technology that underpins plant biotechnology was developed in the public sector in the USA, Europe and Asia and plant biotechnology continues to attract considerable government funding for research all around the world, with China probably the heaviest investor. The commercial development of GM crops in the west, however, is dominated by six multinational companies: BASF, Bayer, Dow, DuPont, Monsanto and Syngenta. The domination of any industry by a small number of companies is always a concern, but it is hardly unique to plant biotechnology. Ironically, although the multinationals have retreated from Europe they will be able to return later if public opinion changes. The small and medium-sized home-grown companies and start-ups that might have competed with them, most of which were in the UK, did not survive. Furthermore, the expense of developing new GM varieties, much of it associated with negotiating the regulatory system, means that only large companies have the resources to do it.

5.17 The StarLink and ProdiGene Affairs

StarLink was a trade name given to several GM maize varieties produced by Aventis that, like those varieties discussed in Chapter 3, had been engineered to be resistant to insects by the insertion of a *Cry* gene from *Bacillus thuringiensis*. However, unlike the successful GM maize varieties grown for human consumption, StarLink contained the *Cry9C* version of the gene instead of the *Cry1A* version. It also contained a gene imparting tolerance of the herbicide gluphosinate.

Problems arose with StarLink because, unlike the product of the *Cry1A* gene, that of the *Cry9C* gene does not break down easily in the human digestive system and is relatively heat resistant. For this reason, StarLink had never been approved for human consumption. However, in 1998 the Environmental Protection Agency (EPA) of the USA approved StarLink for commercial growing as an animal feed. It is not clear to what extent the other federal agencies involved in GM crop assessment, the Food and Drug Administration (FDA) and the Animal and Plant Health Information Service (APHIS) were consulted on this decision. The EPA set a zero-tolerance level for the use of StarLink in human food.

Maize is an out-breeder and some cross-pollination between StarLink and maize varieties destined for human consumption was inevitable. This should undoubtedly have been foreseen by the EPA and Aventis but apparently it was not. In mid-September 2000, the Washington Post reported that StarLink maize had been detected in processed maize snack foods already on the shelf in grocery stores. The products were recalled, the FDA became involved and in the end 300 maize products had to be taken off the shelf. Products were also recalled in Japan and Korea.

Aventis instructed its seed distributors in the USA to stop sales of StarLink seed corn for planting in 2001 and voluntarily cancelled its EPA registration for StarLink. However, Aventis predicted that it would take four years for StarLink to clear the food chain entirely. The company agreed to buy back the entire StarLink crop of 2000 at a premium price, something that must have cost them upwards of $100 million. The 1999 crop was already beyond recall. The incident left Aventis vulnerable and it was subsequently acquired by Bayer.

Another incident involving the accidental mixing of a non-food GM crop with food intended for human use occurred in November 2002, although in this case the crop in question did not go beyond the farm. The biotechnology company involved was ProdiGene, a company that specialises in developing GM crops to produce pharmaceutical products (see Chapter 3). Clearly, these crops are not meant to enter the food chain.

In 2001, ProdiGene contracted a Nebraska farmer to grow one of its experimental GM maize varieties. However, after the crop had been harvested the land was not treated any differently to land that had an ordinary crop growing on it. The next year the farmer planted soybean destined for human consumption on the same land. Inevitably, a small number of maize plants arising from spilled seed grew amongst the soybeans and a tiny amount (65 g) of corn stalks were discovered in the harvested soybean seed. The Food and Drug Administration ordered the soybean crop, worth $2.7 million, to be destroyed.

Such a release in the UK would have to be monitored regularly during all stages of development. Post-harvest, the plants and/or seed would be stored in a secure area; all ventilation would be enclosed and all dust from extraction units or from drying and storage areas would be inactivated by autoclaving (heating under pressure). All straw and similar material would be returned to the release site and burnt prior to cultivation into the soil. After harvest, the site would be ploughed and monitored. Irrigation would be used to encourage germination of volunteers and any re-growth or volunteers would be removed by spraying with an appropriate total herbicide. No crop would be sown on the site at least for the following year, probably the one after as well. The release site would be monitored for two years and any re-growth or volunteers would be removed by spraying with an appropriate herbicide or by cultivation.

All this would apply to a crop designed eventually for food use, never mind one making a pharmaceutical product. Clearly, similar practices should have been used in this case. However, the case does not change the fact that the use of GM plants to produce pharmaceutical products has enormous potential.

The StarLink and ProdiGene incidents should not have been allowed to happen and they highlight the need to segregate crops designed for food and non-food uses. Regulatory authorities will have to decide whether out-breeding species such as maize are suitable plants for the purpose of making pharmaceutical and other non-food products. In fact, earlier in 2002 the United States Department of Agriculture ordered a different ProdiGene GM maize trial in Iowa to be destroyed to prevent cross-pollination with food maize growing nearby. Decisions such as that should be made before the crop is planted and some crops for which there is zero tolerance of presence in the food chain will have to be contained under glass. Nevertheless, it should be remembered that this problem is not unique to GM crops. In the United Kingdom and elsewhere, for example, there is long experience of growing non-GM oilseed rape varieties to produce industrial oils. These oils are not suitable for human consumption and there is a strict tolerance level enforced for the non-food rapeseed in the food rapeseed harvest.

5.18 The *Cauliflower mosaic virus* 35S RNA Gene Promoter

Another scare-story that has arisen in the GM debate in Europe is that the use of the *Cauliflower mosaic virus* 35S RNA gene promoter (usually called the CaMV 35S promoter, see Chapter 2) in plant biotechnology represents some sort of human health risk. *Cauliflower mosaic virus* infects cauliflowers (hence its name) and other plants in the *Brassica* (cabbage) family. It does not infect animals, including humans, at all, but it infests most of the *Brassica* crop of the United Kingdom to some extent, and probably always has done. It is not, therefore, a good candidate for a food scare.

The CaMV 35S gene promoter is only a small part of the viral genome. It is used in plant biotechnology where a gene introduced by genetic modification is required to be active everywhere in the plant. The supposed risk associated with it is that some sort of recombination could occur between this promoter and an animal virus, producing a new 'supervirus'. It is not clear exactly how this is supposed to happen, why it would be more likely to occur when the CaMV 35S promoter is integrated into a plant genome as opposed to a viral genome, or why, even if it did occur, the resulting virus would be particularly dangerous.

5.19 Implications for Developing Countries

The effect that GM technology will have on developing countries is a question that has worried many people and there has certainly been concern in developing countries either that they would become reliant on first-world companies for their seed or, conversely, that they would be denied access to GM technology altogether. These worries still persist.

It is important to note the potential benefits of GM technology for developing countries, as well as the problems. The potential of high vitamin A GM crops such as Golden Rice, virus resistant and high yielding crops, and crops making vaccines to relieve famine, malnutrition and disease in developing countries is obvious. One of the more satisfying developments in the GM crop story of recent years is that developing countries are making their own decisions on the matter (Table 3.1), much to the chagrin of western anti-GM pressure groups.

Developing countries are not all the same, of course; there is a big difference between countries such as India and China that have a large science base of their own and poorer countries that do not. Both India and China are investing heavily in plant biotechnology and China has developed several GM crop varieties for its own

Table 5.1 GM crop varieties developed in China.

Species	Trait
Cotton	Insect resistant
Petunia	Colour altered
Tomato	Virus resistant
Tomato	Increased shelf life
Sweet Pepper	Virus resistant
Papaya	Virus resistant

farmers (Table 5.1). In India, there was strong resistance to the cultivation of GM crops until only a few years ago. This resistance has broken down due to the success of insect-resistant (Bt) cotton, the first variety of which to be made available to farmers was introduced in 2002. This variety was developed by Mahyco (Maharashtra Hybrid Seed Company) using the *Cry1Ac* gene (Chapter 3), licensed from Monsanto. The area planted in 2002 was only 29,000 hectares but, by 2009, 5.6 million Indian farmers were planting 8.4 million hectares of Bt cotton, equivalent to 87% of the national cotton crop, an astonishing adoption rate (data from the International Service for the Acquisition of Agri-biotechnology Applications (ISAAA)). Indian farmers report yield increases, reduced insecticide use and an increase in net profit as the benefits of using GM cotton and many other seed companies now market their own GM varieties under licence. These varieties include some with multiple *Cry* genes and varieties from public as well as private sector plant breeding.

GM insect-resistant cotton has also been adopted in Burkina Faso, GM maize is grown in Egypt and both GM maize and cotton are grown extensively in South Africa. Elsewhere in Africa, arguably the continent with most need to exploit innovations in plant breeding, GM crops are still not being made available to farmers. An example of an African scientist who advocates strongly that the potential of GM crops should at least be explored is Florence Wambugu, the director of the African Centre of the

ISAAA in Kenya. Wambugu, who studied for her PhD at the University of Bath in the UK, stated in her book, *Modifying Africa: How Biotechnology can Benefit the Poor and Hungry* (F.M. Wambugu, 2001) that 'Having missed the Green Revolution, African countries know they cannot afford to pass up another opportunity to stimulate overall economic development through developing their agriculture'. However, African governments appear to be strongly influenced by opinion in Europe, at least on this issue.

An example of this was the controversy over the supply of food aid to Southern Africa in 2002. At first, encouraged by European activists, Malawi, Swaziland, Lesotho, Zimbabwe, Mozambique and Zambia all rejected food aid from the USA that contained GM maize, despite being threatened with famine as the result of a drought. As it became clear that people were going to die without the food aid the activists became silent, fearing a backlash. The bogus health issue receded and the problem became one of future trade with Europe. The African governments feared that some of the imported seed would mix with seed being planted for future harvests and this would jeopardise trade with Europe once the drought was over. At first, the EU refused to reassure the Africans that GM crops were acceptable, afraid that this would threaten their trade bans on the importation of some GM foods from the USA. However, it eventually relented and gave the reassurances that were asked for. Most of the countries affected then agreed to accept the food aid but Zambia, despite the dire need of its people for food, would not.

5.20 'Terminator' Technology

'Terminator' technology is a term that has come to be used to describe the production of crop varieties that produce infertile seeds, or 'suicide seeds' as they are sometimes referred to. Pressure groups that oppose the use of genetic modification in plant breeding have long sought to associate GM crops with 'terminator' technology, indeed the term 'terminator' and the prospect of biotechnology companies using such a technology to force farmers, particularly those in developing countries, to buy seed from them every year has been one of the most effective weapons in the anti-GM campaign. This is astounding given that none of the GM varieties in commercial use for food production produce infertile seed.

It is certainly possible to render the seeds of plants infertile using genetic modification; it can and has been done using non-GM methods as well. Similarly, both GM and non-GM methods can be used to render pollen infertile; indeed, it is an established technique in plant breeding for the production of hybrid seed (see next section), although the hybrid seed that is produced from such plants to be sold to farmers has its fertility fully restored. There may well be applications for the commercial use of sterile plants in specialist, small-scale applications such as the production of vaccines or other pharmaceuticals where it is important to prevent crossing and mixing with crops being grown for food. I am not aware of any breeder developing a variety, GM or otherwise, with sterile seeds for everyday use, and it is difficult to envisage why farmers would buy such a variety if it were launched.

5.21 Unintentional Releases

On 17 April 2000, Advanta Seeds told the UK government that GM-contaminated oilseed rape seed had been sold to farmers across the UK. The incident arose from inadvertent crossing of oilseed rape being grown for seed in Canada. The seed was to be a non-GM hybrid, and to ensure that all of the seed would be hybrid the seed-producing line was male sterile, producing no pollen. Pollen was to be provided by the other would-be parental line which was planted alongside. This technique is used widely in plant breeding and relies on the pollinating parent producing pollen at exactly the right time to coincide with the seed-producing parent becoming receptive. Otherwise, in the absence of any pollen from the intended male donor, the seed-producing plants are pollinated by whatever compatible pollen is blowing in the wind. That is what happened in this case. These things happen in plant breeding but, on this occasion, some of the pollen in the wind was from a GM glyphosate-tolerant oilseed rape being grown on a nearby farm and the seed was sold in Europe, specifically the UK, Sweden, France and Germany. The GM variety had not been approved for cultivation in Europe and an almighty hoo-hah ensued, despite the fact that only 1–3% of the seed was GM and the GM oilseed rape variety in question was grown widely in Canada (and still is), was approved for food use, and oil from it had been consumed by probably millions of people.

Several thousand hectares had been sown with the 'contaminated' seed in 1999 and seed had been harvested and processed and the oil and meal consumed before the incident came to light; more was planted in 2000 and was in the field when the problem was recognised. However, not surprisingly, no ill effects were reported and there is no evidence of persistence of the GM trait in the UK or elsewhere in Europe.

Europe's strict laws governing the use of GM seeds and products make it almost inevitable that this sort of incident develops into a 'crisis'. Similar problems arise when crop products imported for food use are found to contain GM material that is not approved. In 2005, for example, it was reported that seed of an authorised Syngenta GM maize variety, Bt11, contained a small amount of seed from another transgenic line, Bt10. Bt10 contained the same Bt gene for insect resistance as Bt11, but the gene was expressed at a much lower level, making the protection against insect pests ineffective. Development of Bt10 had been stopped for this reason, but some Bt10 had persisted in Bt11 seed production for several years. The Bt11 variety was being grown widely in the USA and grain was being imported into Europe for food and feed use. The imported grain potentially contained small amounts of the unapproved variety, Bt10.

This should not have happened and American farmers who bought Bt11 seed would not have got the protection against insects that they paid for in a small proportion of their maize plants. However, there was no suggestion that the problem represented a threat to human health or the environment. Nevertheless, in 2005 the European Union required Member States to sample shipments of raw or minimally processed maize products coming from the USA at ports, feed mills and distributors' premises. The UK's Food Standards Agency collected samples over a four-week period in September and October of 2005 but was unable to detect any Bt10 grain.

5.22 Asynchronous Approvals

Another increasing difficulty for the European Union is the speed of approval and commercialisation of GM varieties outside Europe compared with the time it takes for new GM varieties to get through the European approval process. This leads to the problem of 'asynchronous approvals', with farmers outside Europe growing varieties that cannot yet be imported for food or feed use. Europe is not self-sufficient in food and feed production and, for example, has to import soybean from elsewhere for the production of animal feed. In 2007, US authorities licensed a new GM soybean variety, Roundup-Ready II, but this was not authorised in Europe until December 2009, and that only occurred as fast as it did because the European animal feed industry faced a shortfall in supply.

5.23 The United Kingdom Farm-Scale Evaluations

In 2000 the UK government was faced with the prospect of the first GM crops becoming available for commercial cultivation in the United Kingdom while the public remained fearful and hostile. In an attempt to instil confidence and to satisfy the demand that the potential effects of GM crops on the environment should be tested more thoroughly before they were grown widely the government struck a deal with the biotechnology industry. This was called the SCIMAC (**S**upply **C**hain **I**nitiative for **M**odified **A**gricultural **C**rops) agreement. The industry undertook not to commercialise GM crops before 2003. In the meantime, farm-scale evaluations (field trials) would be carried out on the first generation of GM crops proposed for use in the UK in order to determine what effect they would have on the environment. The trials management committee included representatives from English Nature, the government agency responsible for nature conservation in England, and the Royal Society for the Protection of Birds (RSPB).

The government stated that commercialisation after the farm-scale trials were completed would not proceed if the GM crops were shown to have an adverse effect on biodiversity. This had never before been asked of an innovation in agriculture and is still not asked of new non-GM varieties or of new or old farming methods.

The GM crops in the trials were gluphosinate-tolerant (Liberty Link) oilseed rape and fodder maize from Bayer, and glyphosate-tolerant (Roundup-Ready) sugar and forage beet from Monsanto. These varieties were potentially close to commercialisation in the UK and the focus on them made sense from that point of view. However, it meant that the trials concerned only herbicide-tolerant crops used in conjunction with the appropriate herbicide, and did not enable a general assessment of GM crops or genetic modification to be made. Unfortunately the trials also followed narrow crop management systems devised specifically for weed control. The results of the trials were announced in 2003 and showed that although biodiversity (that is the number and variety of weeds and insects) in the GM maize was higher than in the non-GM maize, for the other crops the GM varieties scored lower for biodiversity. Ironically, some maize farmers who took part in the trials were unimpressed with the weed control provided by the GM maize in combination with gluphosinate: farmers think their primary role is food production and weeds impact on their yield.

The UK media geared up for the release of the Farm Scale Evaluation results with a steady stream of anti-GM headlines. After the report was published, many newspapers stated that the results represented 'the end of GM in the UK'. The UK government had put itself in a position where it would have been politically extremely difficult to allow commercialisation of the GM crops that had a bad biodiversity report in the trials. However, it was even unwilling to allow commercialisation of Bayer's GM maize variety, stating that it would consider the issue of segregation and co-existence with non-GM crops first and giving no deadline for a decision. Bayer then announced that they would not proceed with commercialisation.

The UK Farm Scale Evaluations programme provided a lot of information on the insects and weeds in UK farmland. They did very little to inform the debate on GM crop use in the UK. The scientists involved did a very professional job in the execution of the experiment and the reporting of the data, but failed to ensure that the media were aware of the limitations of the study. Indeed, they were hopelessly naïve in expecting the results of the study to be reported in a balanced way without their being much more pro-active in their dealings with the media. There is no doubt that the process put any prospect of GM crop cultivation in the UK back by years.

5.24 Conclusions

Genetic modification is now an established technique in plant breeding in many parts of the world. While not being a panacea, or replacing every other method in plant breeding, it does enable plant breeders to introduce some traits into crop plants that would not be possible by other methods. GM crops now represent almost 10% of world agriculture by area, and are being used in developed and developing countries. Farmers who use them report one or more of greater convenience, greater flexibility, simpler crop rotation, reduced spending on agrochemicals, greater yields or higher prices and increased profitability at the farm gate as the benefits.

That does not mean that every farmer who has the option to use GM varieties does so. GM varieties enter a very competitive marketplace and suppliers offering alternatives reduce their prices or offer other incentives to make their own products competitive. Nor is it the purpose of this book to tell farmers what varieties, GM or otherwise, they should or should not grow, only that UK and European farmers should be given the choice of growing GM crops that have come through the testing and regulatory processes. Otherwise they will find it increasingly difficult to compete with overseas farmers.

The delay in allowing plant biotechnology to develop in Europe has already damaged the European plant biotechnology industry significantly and is putting European agriculture at an increasing competitive disadvantage. The consequences of losing the debate, which is where we are at the time of writing, can be summarised as:

- Currently no GM crops are being developed specifically for European conditions or the European market.
- The agricultural biotechnology industry is currently focused on obtaining consent for import of more GM crop products into Europe, rather than for permission to market GM crop seed for cultivation in Europe.
- European farmers are increasingly disadvantaged in a competitive global market, competing with GM crops but unable to use them.
- The agricultural biotechnology multinationals have closed or down-sized their biotechnology research and development operations in Europe.
- The small agricultural biotechnology start-up industries that appeared in Europe in the 1980s and 1990s have largely been lost.
- The USA, China, India, Brazil, Argentina and others have a huge lead over Europe in a key twenty-first century technology.

The GM crop issue first came to the attention of the UK public in 1996, already a decade and a half ago. The UK appears to be no closer to commercial GM crop cultivation now than it was then. However, the world has changed in that time. The 1990s were in the era of crop surpluses in Europe, with plentiful cheap food and still relatively cheap energy. With news stories of European farmers being paid to destroy produce being a regular feature, increasing crop yield was not even seen as a worthwhile target for plant scientists. Food shortages elsewhere in the world were talked of as issues of distribution rather than production. Warnings that a small surplus of food was much better than too little, and that the situation would not last, were ignored.

The issue of crop yield and the need to increase food production crashed back on to the global agenda in the second half of the last decade as food prices soared. The price of wheat on the London International Financial Futures and Options Exchange (LIFFE), for example, rose to a record £198 in 2008 as a result of increasing demand and severe drought in Australia. The Australian drought could have been seen as a one-off event, but there was concern that it was the sort of extreme weather event that would occur with increasing regularity through the twenty-first century as a result of climate change. Increasing demand was also something that could not be ignored. World population passed six billion in 1999. In 2011 it will pass seven billion and it is predicted to peak at nine billion around the middle of this century before starting to decline slowly. In addition, greater prosperity in highly populated, economically rapidly emerging countries such as China and India is leading understandably to a demand for a better diet, in particular a demand and ability to pay for more meat. On top of that, crops are now being used for fuel as well as food. In 2010, 33% (over 100 million tonnes) of the US maize harvest was used for bioethanol production and 25% of UK wheat could be used for biofuel by 2015 (Chapter 3).

Farmers responded to the increase in food prices in 2007–2008 by increasing production, as they always have done, and prices fell back. However, 2010 saw another major commodity crop exporter, Russia, suffering from drought and prices climbed again. In December 2010, the LIFFE price hit a new record of £200 per tonne. Prices will no doubt continue to fluctuate, but arguably the era of cheap food and global food surpluses has ended.

Agriculture will face several difficult challenges in the coming decades: as well as population growth and climate change there will be problems of fresh water supply, peak oil (the point when the maximum rate of global petroleum extraction is reached, after which the rate of production enters terminal decline), competition for land use, the need for wildlife conservation, soil erosion, salination and pollution. Furthermore, while agriculture will be affected by climate change, it will also be expected to contribute to efforts to cut greenhouse gas emissions in order to avoid the worst-case scenario predictions of global temperature increases. In the UK, for example, the Climate Change Act (2008) commits the country to an 80% reduction in greenhouse gas emissions by 2050 across all sectors of the economy, and farming is responsible for about 7.4% of total UK emissions. If agriculture is going to meet these challenges, plant breeding will have to play its part and plant breeders will have to be able to use every tool that is available, including biotechnology.

INDEX

α-amylase 82, 104
α-amylase/trypsin inhibitors 80, 82, 104
α-linolenic acid (ALA) 80
ABA response element binding protein (AREBP) 92
abscisic acid (ABA) 91, 92
acetolactate synthase (*ALS*) 38
acetyl-CoA 70
Achromobacter 59
acifluorfen 61
actin 45
adenine 5, 7
administration 25, 75, 100, 124, 140, 155, 156
Advanta 161
Advisory Committee for Animal Feedingstuffs (ACAF) 118
Advisory Committee for Releases into the Environment (ACRE) 111, 118
Advisory Committee on Genetic Modification (ACGM) 107
Advisory Committee on Novel Foods and Processes (ACNFP) 115, 118, 120, 144, 150
Aequorea victoria 40, 41
Agritope, Inc. 57
Agrobacterium rhizogenes 27
Agrobacterium tumefaciens 26–31, 34, 38, 42, 59
agrochemicals 17, 69, 165
Alabama 66
albumins 103, 135
aleurone 41
alfalfa 99, 100
Alfalfa mosaic virus (AlMV) 100
Alkaptonuria 3
allele 15
allergenicity 104, 105, 110, 115, 117, 134, 135
allergens 103, 117, 134, 135
Amflora 81, 118
amino acid 8, 9, 27, 37, 38, 59, 76, 83, 96, 135, 140
aminocyclopropane-1-carboxylic acid (ACC) deaminase 57
aminocyclopropane-1-carboxylicacid (ACC) synthase 57
aminoglycosides 35
ampicillin 151
amylopectin 80, 81
amylose 80, 81

anaphylaxis 104
Animal and Plant Health Inspection Service (APHIS) 124, 147, 155
animal feed 18, 19, 40, 61, 81, 83, 118, 121–123, 135, 137, 138, 145, 155, 163
animal feeding studies 137
anther 43, 44
antibiotic resistance 28, 35, 36, 150, 151
antibody 101, 102
antifreeze gene 137
antisense 48, 49, 104
apple 104
Arabidopsis 10, 11, 15, 26, 34, 78, 79, 95, 97, 99
Arabidopsis genome 12, 24
arachidonic acid (AA) 71
Argentina 53, 54, 60, 61, 119, 166
Argonaute 49
Arthrobacter globiformis 97
Arthrobacter panescens 97
Asia 14, 119, 128, 154
asynchronous approvals 163
Australia 54, 82, 91, 119, 149, 167
Austria 3, 118, 119, 145
auxin 30
Aventis 62, 64, 127, 155, 156
Avery, Oswald 4
avirulence (*Avr*) genes 88
avocado 104

β-carotene 85, 86
β-glucanase 90
β-glucuronidase 38, 40–43
Bacillus amyloliquefaciens 140
Bacillus thuringiensis 65, 67, 155
back-crossing 18, 89
Baker's asthma 104
BAR gene 36
barley 20, 32, 46, 47, 83, 104
base pair 7, 8, 10, 11, 24, 28, 44, 46, 47
BASF 79, 81, 89, 118, 154
Baulcombe, David 49
Bayer 62, 64, 127, 154, 156, 164, 165
Beadle, George 4
bean 20, 28
Bechtold, Nicole 34
bentazon 61
Berg, Paul 24
betaine aldehyde dehydrogenase (BADH) 96, 97
Bevan, Michael 28
bialaphos 36, 37
binary vector 28
biodiesel 75, 77
biodiversity 111, 153, 164
bioethanol 19, 81–83, 167
biolistics 32
biological containment 108
biological control 65
bioluminescence 40
biopharming 97, 103
bio-piracy 153
birch 104
Blumwald, Eduardo 95
borage 74, 78
bovine spongiform encephalopathy (BSE) 130
Boyer, Herbert 25
BP 81
Brachypodium 10
Bragg, William 6
Brake, Denise 145
Brassica oleracea 14, 15

Brazil 53, 54, 61, 119, 134, 166
brazil nut 134
bread-making 13, 14, 19
British Medical Association 150
British sugar 81
broccoli 14
bromoxynil 63, 64
Brown streak virus 70
brussels sprouts 14, 146
Bt 65–69, 119, 124, 141, 147–149, 159, 162
Burkina Faso 54, 159

cabbage 14, 15, 44, 103, 146, 157
Calgene 56, 76, 124, 125
Californian Bay Plant 76
callus 27–29, 31, 33
Canada 20, 53, 54, 63, 64, 75, 119, 137, 146, 153, 161, 162
canine parvovirus 100
canola – see oilseed rape
capric acid 73
caprylic acid 73
carbohydrate metabolism 92
carbon dioxide 81
carcinoembryonic antigen (CEA) 102
cardiovascular disease 79
Cargill 83
carrot 104
cassava 68, 70, 80, 87
Cassava mosaic virus 68, 70
catalase 93
catering outlets 121
cauliflower 14, 43, 45, 157
Cauliflower mosaic virus 43, 45, 157
celery 104
cell wall 26, 30, 32, 55, 56, 90
cellulase 30, 56
cellulose 56
cereals 10, 17, 31, 32, 35, 37, 41, 80, 84, 93, 104
chacocine 89
Chang, Annie 25
chaperones 93, 94
charlock 146
chemical barrier 108
chemical mutagenesis 19, 51
cherries 104
chestnut 104
Chilton 26, 28
chimaeric gene 42, 46, 48
chimaeric virus particles (CVP) 100
China 53–55, 69, 128, 153, 154, 158, 159, 166, 167
chitinase 90
chlorimuron 61
cholera 99, 151
cholesterol 74, 77
choline dehydrogenase (CDH) 97
choline mono-oxygenase (CMO) 96
choline oxidase (COD/COX) 97
chorismate 38, 59
chromosome 10, 19, 89
chromosome doubling 19
Chua, Nam-Hai 43
climate change 90, 167, 168
coat protein 69, 100
coconut oil 76
coding region 8, 9, 42, 47
co-existence 148, 165
Cohen, Stanley 25
colchicine 19
cold 90, 92, 93

Colorado beetle 65, 67
compatible solutes 95, 96
Conservation tillage 61, 62
constitutive gene expression 9
constitutive promoter 43
consumers 77, 84, 87, 103, 105, 120, 123, 128–130, 133, 137, 148, 150
containment 29, 107–111, 114
corn – see maize
corn borer 66, 68, 83, 142
corn rootworm 83
co-suppression 48, 49
cotton 53, 62, 63, 65, 66, 122, 123, 148, 159
Council of Ministers 118
Cowpea mosaic virus (CPMV) 100
Crick, Francis 6
crop rotation 58, 61, 165
cross-protection 69
crown gall 26, 27
crucifers 103
Cry (gene or protein) 65, 67, 147, 155, 159
cup-shaped cotyledon (CUC) 92
Cuvier, Georges 2
cyanogenic glycosides 17, 133
cytokinin 30
cytomegalovirus 99
cytosine 5–7

daffodil 85, 87
Daily Mail 131, 132, 145
Darwin, Charles 1
dehydration-responsive element binding protein (DREB) 92
delta-5 desaturase 78
Denko, Showa 140
Denmark 118
deoxyribonucleic acid 4
deoxyribose 5, 7
Department for Environment, Food and Rural Affairs (DEFRA) 111
Department of the Environment in Northern Ireland 111
developing countries 70, 80, 85, 99, 128, 158, 161, 165
developmentally regulated promoter 43
diabetes 74, 79, 97, 98
dicotyledonous plants 31, 35, 45, 63
dihydrodipicolinate synthase (DHDPS) 83
diploid 14, 89
direct gene transfer 30–32, 34
Directive 2001/18/EC 111
disease resistance 18, 88
DNA 4–11, 19, 23–28, 30–34, 44–46, 57, 92, 100, 112, 120, 121
DNA cloning 26
DNA ligase 23
DNA-mediated gene transfer 30
DNA Plant Technologies 57
DNA polymerase 23, 25
docosahexaenoic acid (DHA) 78–80
dog 100
double helix 7
Dow 154
drought 82, 90–95, 160, 167
dry grind 82
DuPont 76, 79, 81, 127, 137, 154
dwarfing genes 17

5-Enolpyruvylshikimate phosphate (EPSP) 38, 39, 59

5-Enolpyruvylshikimate phosphate synthase (EPSPS) 37–39, 59, 60, 62
E box 47
edible vaccines 101
egg sacs 43, 44
eicosanoid 74, 78
eicosapentanoic acid (EPA) 66, 78–80, 124, 147, 148, 155, 156
electroporation 31, 32
Ellis, Jeff 34
embryo rescue 18
Endless Summer 57
endosperm 41–43, 46, 47, 85
English Nature 164
enhancer 47
Environmental Protection Act 111
Environmental Protection Agency (EPA) 66, 78, 79, 124, 147, 148, 155, 156
eosinophilia-myalgia syndrome (EMS) 140
Ermakova, Irina 144
erucamide 75
erucic acid 17, 18, 75
Erwinia uredovora 85
Escherichia coli 25, 26, 28, 35, 99, 151
Essay on the Principle of Population 16
essential amino acids 38, 59, 76
essential fatty acids (EFAs) 77
ethylene 55–57
Europe 52, 67, 68, 81, 118–121, 127–130, 137, 166
European Commission 67, 68, 118, 119, 120, 145
European corn borer 83
European Food Safety Authority (EFSA) 118, 144, 145
European Union (EU) 67, 68, 81, 83, 111, 118, 119, 122, 133, 150, 160, 163
evening primrose 74, 78
Evenson, Donald 145
evolution 1, 2, 4, 11, 17, 88, 160
explants 28, 32, 34
extinction 1, 2

farm-scale evaluations 163
farmscale trials 164
fatty acid methyl ester (FAME) 77
fatty acids 70–72, 74, 75, 77, 78, 80, 123
FDA – see Food and Drug
Feathery mottle virus 68, 70
fermentation 122
fertilizer, 12, 44
field trial/release 111
fish oil 78–80, 123
Flavr Savr 56, 124, 125, 139
flax 20
floral dip transformation 34
flour 82, 104
fodder beet 62, 63
folic acid 84
food allergy 103, 135
Food and Drug Administration 25, 75, 124, 140, 155, 156
food safety 17, 20, 107, 118, 138, 144
Food Standards Agency (FSA) 103, 115, 118, 122, 123, 128–130, 145, 163
foot and mouth disease virus (FMDV) 99

forced cross 18, 146
Forde, Brian 47
Fraley, Robert 28
France 68, 91, 118, 119, 145, 162
Frankenstein food 132, 136
Franklin, Rosalind 6, 7
freezing 9, 92
French bean 28
fumicosin 67
fungal resistance 88, 90
fungicidal proteins 90

galactose 56
galacturonic acid 56
gamma linolenic acid (GLA) 73–75, 78, 89, 108, 117, 157, 164
gamma rays 19
Garrod, Sir Archibald 3
gene definition 8, 9
gene expression 9, 47, 49, 117, 134
gene flow 136, 146, 148
gene number 11
gene over-expression 47
gene silencing 47–49
gene structure 25
gene terminator 9
General Agreement on Trade and Tariffs (GATT) 153
genetic change 11, 20, 51
genetic diversity 153, 154
genetic engineering 21, 131
Genetic Modification Safety Committee 110
genetically modified organism (GMO) 107, 108, 110, 111, 121, 122
geneticin 35
genetics 1, 3, 10, 14, 24, 28, 88, 95, 115
genome 1, 10, 11, 14, 21, 26, 29, 32, 45, 62, 100, 112, 151, 158
genome duplication 14
geranylgeranyl diphosphate 85
gibberellin 17
Gilbert, Walter 25
glucan 56, 82
gluco-amylase 82
glucose 56
glucosinolates 17, 18, 75, 134
Glu-D1x 42, 43
gluphosinate (glufosinate) 37, 62, 63, 119, 137, 155, 164
glutamic acid 96
glycinebetaine 95–97
glycoalkaloids 17, 89, 133
glycophytes 95
glycoside 17, 35, 133
glyphosate 37–39, 58–64, 67, 83, 137, 154, 162, 164
glyphosate oxidoreductase (GOX) 37, 38, 62
GM enzymes 122
GM Food and Feed Regulations (EC) 118
GM food safety 138
GM-free 149
GMO deliberate release regulations 111
Golden Promise 20
Golden Rice 84, 86, 87, 128, 158
Gordon, Milton 26
grape 10
Greece 118, 119
green fluorescent protein (GFP) 38, 40, 41
Green Revolution 17, 160

Greenpeace 86, 87, 131
Grierson, Don 48, 56
guanine 5–7
GUS – see *UidA*

H^+-ATPase 95
hairpin RNA (hpRNA) 49
hairy root disease 27
Hall, Tim 28
halophyte 95
Hamilton, Andrew 49
Hawaii 69
Health and Safety Executive (HSE) 107, 109, 111
heat 90, 91, 93–95, 99, 108, 155
heat shock factor (HSF) 93, 94
heat shock protein (HSP) 93, 94
Helling, Robert 25
hemicellulase 30
hepatitis B 98, 99
herbicide 17, 35–38, 43, 55, 57–64, 108, 110, 114, 131, 136, 137, 139, 144, 146, 147, 154–157, 164
herbicide tolerance 35–37, 43, 55, 57, 62, 63, 136, 137, 146, 154
herbivore 17, 89
heredity 8
herpes simplex virus (HSV) 102
hexaploid 14, 19
horizontal gene transfer 110, 151
hormone 17, 27–30, 33, 55, 72, 91
Horsch, Robert 28
housekeeping gene 9
HPH/HPT 35
human genome 10, 11, 26
hybridisation 14, 25
hygromycin phosphotransferase (*APH-IV/HPH/HPT*) 35
hypersensitive response (HR) 88

imazaquin 61
imazethapyr 61
imidazolinone 137
Imperial Chemical Industries (ICI) 48
India 53, 54, 119, 138, 146, 153, 158, 159, 166, 167
inducible promoter 45
inheritance 3, 4, 15
insect resistance 43, 55, 64, 69, 137, 147, 162
insecticide 66, 67, 124, 142, 147, 159
insulin 25, 97, 98
International Convention to Protect New Varieties of Plants (UPOV) 152
International Rice Research Institute 87
International Service for the Acquisition of Agri-biotechnology Applications (ISAAA) 159
Irish potato famine 88
iron 85, 86
irradiation mutant 20
irrigation 94, 114, 156
Italy 14, 20, 118

Japan 87, 128, 140, 155
junk DNA 11

kale 14
kanamycin 35, 36, 124, 150
Kenya 70, 160
keratomalacia 85
Klebsiella pneumoniae 63, 151

Korea 155
Kornberg, Arthur 23

labelling 120–122, 124, 125
lacewing 141, 142
ladybird 141, 142
Lamarck, Jean-Baptiste 2
late blight disease 88
Lauric acid 72, 76, 77
Lawes, John 12
lectin 142
legumes 103
Lesotho 160
leukotriene 74
Liberty 62, 63, 164
Liberty Link 164
Linnaeus 12
linoleic acid 74
lipid transfer proteins 104
Lol p5 allergen 104
London International Financial Futures and Options Exchange (LIFFE) 91, 167
long chain polyunsaturated fatty acids (LC-PUFAs) 74, 78, 79
Losey, John 141
lotus 10
luciferase (*Lux*) 38, 40, 46
Luxembourg 118, 119
lycopene β-cyclase 85
lysine 83

Ma, Julian 102
MacLeod, Colin 4
Maize 10, 31, 32, 38, 45, 46, 53, 61–63, 65–67, 73, 74, 81–83, 87, 92, 94, 96, 103, 104, 120, 122, 123, 137, 141, 142, 145, 146, 151, 155–157, 159, 160, 162–164, 167
Malawi 160
Malthus, Reverend Thomas 16
marine algae 79, 80, 123
Mason, Hugh 99
mavera maize 83
McCarty, Maclyn 4
Mendel, Gregor 2, 15
metabolic syndrome 79
metabolite profiling 117
metabolomics 117
methionine 57, 134, 135
metribuzin 61
Mexico 54, 119
microarray 26
Miescher, Friedrich 4
mink 100
monarch butterfly 141
monocotyledonous plants 31, 45
Monsanto 53, 54, 56–58, 67, 69, 76, 77, 79, 83, 94, 127, 154, 159, 164
Mouse hepatitis virus 100
Mozambique 160
Mullis, Kary 25
multinationals 154, 166
mustard 103, 146
mutagen 19
mutagenesis 19, 20, 30, 51, 75, 101, 104, 117, 136, 137
mutation 4, 19, 20, 30, 34, 52
mycotoxins 66
myeloblastosis (MYB) 92
myelocytomatosis (MYC) 92
myristic acid 73

Na^+/H^+ antiporter 95
NAC family 92
Napier, Johnathan 78

National Council for Food and Agricultural Policy (NCFAP) 61
natural selection 1, 2, 11
Nature (Journal) 4, 11, 18, 35, 51, 52, 94, 112, 116, 141, 142, 164
N box 47
neomycin 35, 150
neomycin phosphotransferase 35
Nester, Eugene 26
Neurospora crassa 4
NewLeaf 67, 69
NewLeaf Plus 69
next generation sequencing 10
nitrate 62
nitrilase 63
nitrogen 5, 17, 37, 46, 47
no apical meristem (NAM) 92
Nobel Prize 6, 7, 23–25
nopaline 27, 42
nopaline synthase (*Nos*) 42
North America 20, 55, 67, 75, 137
Norwalk virus 99
NPTII 35
nucleotide 5, 6, 25, 26, 47, 49
nutraceutical 77
nuts 104

oat 32
octopine 27
octoploid 19
oil (industrial) 75, 77, 157
oil (pharmaceutical) 77
oil (plant) 70, 72, 73, 77, 80
oilseed rape 17, 18, 20, 34, 53, 62–64, 73–79, 83, 96, 122, 123, 137, 146, 149, 150, 157, 161, 162, 164
oleic acid 72, 74–77, 125
omega-3 73, 78–80, 123
omega-6 73, 74, 78
one gene, one enzyme hypothesis 4
opine 27
opinion polls 128
oral tolerance 101
organic base 5, 7, 9
organic farming/food 139, 148–150
osmotic stress 92, 94

palm oil 76
palmitic acid 72, 74
papaya 53, 55, 69, 159
Papaya ringspot virus (PRSV) 69
Paraguay 53, 54
paromycin 35
Part B release 118
Part C consent/release 119
particle bombardment 32, 33, 47, 59, 151
particle gun 31, 32
pasta wheat 14
patenting 57, 152, 153
pathogen-related (PR) proteins 104
Pauling, Linus 6
PBI 76
pea 3, 15, 103
pectin 56
pectin methylesterase (PME) 56
pectinase 30
peer review 143–145
Pelletier, Georges 34
pesticide 17, 65, 66, 68, 137, 142, 147–149
petunia 159

pharmaceutical 25, 77, 97, 101, 115, 124, 137, 140, 156, 157, 161
phenylalanine 38, 59
Philippines 54, 87
phosphate group 5, 91
phosphinothricin 37
phosphinothricine acetyl transferase (PAT) 36, 37, 63
phosphoenolpyruvate 38, 59
phospholipid 74
photosynthesis 63, 93
physical containment 108
phytase 86
phytoene desaturase (*crtT*) 85
phytoene synthase (*psy*) 85
pioneer 137
Plant Biotechnology Inc. 102
Plant Breeders' Rights 152
plant breeding 1, 3, 10–13, 15–18, 20, 30, 43, 51, 75, 104, 105, 111, 117, 133–136, 159–161, 165, 168
Plant Genome Systems 62
plant nuclear factor-y (NF-Y) 92
plantibody 102
plasmid 25–28, 151
Plenish 73, 77
Plum potyvirus (PPV) 100
pollen 104, 105, 108, 114, 132, 141, 161, 162
polyethylene glycol 30
polygalacturonase 56
polyol 96
poplar 10
post-transcriptional gene silencing 48
potato 10, 17, 29, 34, 65, 67–70, 80, 81, 88, 89, 99, 100, 118, 133, 142–144, 146
Potato leaf roll virus (PLRV) 69
Potato virus X (PVX) 100
Potrykus, Ingo 85–87, 128
Prince Charles 131, 136
ProdiGene 99, 103, 155–157
prolamin box 47
proline 95–97
proline dehydrogenase 96
promoter 9, 41–49, 59, 60, 152, 157, 158
promoter (CaMV35S) 43–45, 59, 157, 158
prostaglandin 74
protein 4, 6, 8, 9, 17, 26–28, 36–38, 40–42, 45–48, 52, 59, 61, 65–67, 69, 83, 88, 90–93, 95, 97–104, 116, 117, 120, 121, 134–136, 138, 140, 147, 149
protein kinase 91, 92, 95
proteinase inhibitors 17, 134
proteomics 117
protoplasts 30–32, 34
Pseudomonas chlororaphis 57
Pusztai, Árpád 142–144
pyrroline-5-carboxylate reductase (P5CR) 96
pyrroline-5-carboxylate synthase (P5CS) 96

Qualified Majority Voting 118

R genes 88
rabies 99, 100
radiation mutagenesis 30, 136
radish 103, 146
RB gene 89
reactive oxygen species (ROS) 63

recombinant DNA 23–25
refuges 66
regulatory element 46, 47
replicase 69
reporter gene construct 46, 47
restriction endonuclease (restriction enzyme) 24, 25
Ri plasmid 27
ribonucleic acid 4, 9
ribosome 9
ribulose 1,5-bisphosphate carboxylase/oxygenase (Rubisco) 93
rice 10, 11, 14, 15, 20, 31, 32, 34, 45, 53, 63, 84–87, 96, 103, 104, 128, 158
rice genome 11
Richardson, Charles 23
ripening 54–56, 120
risk assessment 110–112, 116, 124
RNA 9, 26, 42, 44, 45, 48, 49, 93, 94, 157
RNA-induced silencing complex (RISC) 49
RNA interference (RNAi) 49
Rogers, Stephen 28
Romania 54, 67
root 27, 30, 33, 43, 44, 80, 91
root formation 30
Rothamsted 47, 78, 79, 117
Roundup/Roundup-Ready 58, 60, 62, 163, 164
Royal Society 143
Royal Society for the Protection of Birds 164
rubisco activase 93
Russia 65, 67, 82, 91, 144, 145, 167
rye 19, 32, 46, 104
ryegrass 105

2S albumins 103, 135
S-adenosyl methionine (SAM) 57
Safeway 120
safflower 74, 97
Sainsbury 49, 120
Sainsbury laboratory 49
salad farmers 65, 66, 147
salt tolerance 94–97
SAM hydrolase 57
Sanger, Fred 25
Schell, Jeff 28
Schilperoort, Rob 28
Schuch, Wolfgang 48
Scientific Advisory Committee on Genetic Modification (Contained Use) (SACGM (CU)) 107
scoreable marker genes 38, 40
seed 3,13, 17–19, 28, 34, 41–43, 46–48, 58, 60, 62, 68, 72, 81, 85, 91, 93, 96, 97, 103, 104, 110, 114, 115, 117, 121, 122, 128, 134, 139, 149, 154–156, 158–162, 166
seed protein 17, 28, 42
segregation 148, 149, 165
selectable marker genes 34, 35, 43, 57, 62
selective breeding 13, 14
sembiosys 97, 98
sethoxydim 61
sexual transfer 110
shelf-life 48
shikimate pathway 38, 39, 59
short interfering RNA (siRNA) 49

signalling pathways 91, 92, 95
silencer 47
silicon carbide fibre vortexing 32
Smith, Hamilton 23
snowdrop 142
Soil Association 131, 139
soil bacteria 59, 150, 151
soil erosion 62, 167
SOS1 95
South Africa 53, 54, 67, 159
Southern blot 25
Southern, Edwin 25
soybean 10, 17, 34, 53, 58–63, 67, 73, 76, 77, 79, 83, 102, 115, 120, 122, 123, 125, 134, 135, 139, 144, 145, 148, 154, 156, 163
Spain 54, 67, 119
Spotted wilt virus 69
squash 53
starch 17, 46, 80–82, 83, 91, 92
starch synthase 81
starflower (borage) 73, 74, 78
StarLink 147, 155–157
stearic acid 73, 74
storage protein 46, 93, 103, 134
Streptomyces hygroscopicus 36, 62
Streptomyces mutans 102
Streptomyces pneumoniae 4
Streptomyces viridochromogenes 37
substantial equivalence 116, 117
subunit vaccine 99
sugar 5, 7, 27, 56, 58, 62, 63, 65, 81, 82, 92, 95, 96, 120, 122, 150, 164
sugar beet 58, 62, 63, 65, 150
sulphur 46
sunflower 28, 73, 74, 78, 104
superbugs 148
superoxide 63, 93
superoxide dismutase 93
superweeds 64, 146
Supply Chain Initiative for Modified Agricultural Crops (SCIMAC) 163
surface antigen 98, 99
Swaziland 160
swede 81, 103, 162
sweet pepper 69, 159
sweet potato 68, 70, 80
syngenta 48, 82, 87, 119, 127, 154, 162

Tatum, Edward 4
T-DNA (transfer DNA) 27
T-DNA borders 28
technology fee 61
terminator technology 160
tetraploid 14, 19, 89
Thermococcales 82
throat cancer 67
thymine 5, 6, 7
Ti (tumour-inducing) plasmid 26
tissue-specific promoter 43, 46, 47
tobacco 36, 53, 66, 97, 99, 100, 102, 103
Tobacco mosaic virus (TMV) 100
tolerance level 149, 155, 157
tomato 48, 53, 55–57, 69, 92, 93, 95–97, 99, 100, 103, 120, 124, 125, 137, 139, 159
Tomato bushy stunt virus (TBSV) 100
tooth decay 102

toxicity 18, 38, 40, 59, 65, 110, 116
transcription 9, 45, 48, 92–95
transcription factor 92–95
transcriptomics 117
transformation 29, 30, 32–34, 151
transient transformation 31, 49
translation 9
transmissible gastroenteritis virus (TGEV) 99
triticale 18, 19, 136
Triticum aestivum 10, 13
Triticum searsii 13
Triticum urartu 13
tropical countries 55, 67, 85
trypsin 103, 104
tryptophan 38, 59, 139, 140
trypzean 103
Tschermak 19
tumour-inducing (TI) plasmid 26
turnip 103
tyrosine 38, 59

ubiquitin (*Ubi*) 45
UidA 38, 40, 41, 42, 43, 44, 46, 47
UK – see United Kingdom
Unilever 127
United Kingdom 16, 58, 75, 99, 103, 107, 111, 115, 120, 123, 128–130, 146, 150, 152, 157, 158, 163
United States 16, 124, 157
United States Department of Agriculture 16, 124, 157
UPOV convention 152
Uruguay, 38
US – see United States
UV radiation 19

vaccines 97–101, 103, 141, 158, 161
vacuolar Na+/H+ antiport 95
vacuole 95
van Montagu, Marc 28
variation 1–3, 11, 14, 18–20, 90
Vibrio harveyi 40
virulence (VIR) genes 26
virus 43, 45, 48, 49, 55, 68–70, 99, 100, 102, 110, 157–159
virus-induced gene silencing 49
virus resistance 55, 68, 69
visible marker genes 41
vistive 77
vitamin A 84–87, 128, 158
vitamin E 84
vitamin K 84
vitamins 84
Vivergo 81

Wallace, Alfred Russell 1
walnut 115
Wambugu, Florence 159
Washington Post 155
Watson, James 6
weed control 57, 58, 60, 61, 139, 164
Weiss, Bernard 23
Westar Oxy-235 64
wheat 10, 11, 13, 14, 16, 19, 20, 31–33, 36, 41–44, 46, 53, 54, 81, 82, 91, 93, 103, 104, 108, 117, 131, 134, 146, 149, 167
white bean 20
Wilkins, Maurice 6

Wilson, A.S. 19
winter flounder 92
World Health Organization (WHO) 85, 115, 133, 137, 138, 150
World Trade Organization (WTO) 119, 153

xerophthalmia 85
X-ray crystallography 6
X-rays 19

yeast 10, 98, 136
yeast genome 10
yield 16, 19, 58, 60, 66, 70, 82, 90, 91, 94, 96, 139, 158, 159, 164–167

Zambia 160
Zeneca 48, 56
zinc finger homeodomain (ZFHD) 92